Research Methodology

— Techniques and Methods —

The Authors

Prof. (Dr.) Sankar Kr Acharya, former Head, Dept. of Agril Extension and Director, Extension Education, Bidhan Chandra Krishi Viswavidyalaya, Mohanpur, WB, born on 6th October, 1960, started his career as Assistant Professor at BCKV in 1988 and has been in teaching, research and extension over 30 years. An erudite teacher as well as an elegant speaker, Prof. Acharya has also been the topper and awarded the certificate of merit in his M. Sc. (Ag) academic career in Agril. Extension. He is internationally acclaimed for his unique research domain of Social Entropy and Energy Metabolism, Social Ecology and Environmental Sociology, Enterprise Ecology Framework, Farm Stewardships: The Transition from Conservation, Technology Socialization Process, and The Social Ecology of Livelihood, which have got tremendous policy as well as scholastic importance**. Research Publication: 227 in National and International Journals. Book Publication: 103 books** authored/co-authored, covering areas Participatory Extension Planning and Management; Value, Extension Strategy for Natural Resource Management; Horticulture Management with Quality Issues, Technology Socialization Process; Fish bio-diversity; Income and Livelihood; Enterprise Management and Entrepreneurship Building;; Social Metabolism, Farm Energy Metabolism, Community Perception on Climate Change, Social Entropy and Chaos,Decision ecology Commission; Marketing and Extension Sub Committee, Govt. of West Bengal; Expert member WWF projects in BTR (Buxa Tiger Reserve Project); Expert member; DFID project on Primary Education (DPEP); REIA, Visited Italy, France, Germany, China, Sri Lanka and Bangladesh. Delivered 17 Key note addresses, 13 guest, invited and lead speakers in International Congress, Chaired 19 Sessions in International conferences, Editors of a score of national and international journals. He has so far been awarded with 13 best paper awards in National and International Conferences Organizing Secretary of two International Conferences, held in the years, 2016 and 2018, Jointly Organized by Department of Agricultural Extension, BCKV, WB and Krishisanskriti, New Delhi. His **Present RG score: 28.64.** He has also been honoured to be selected as the **Convener of Panel (PE-32) entitled The hunger, poverty and silence, of IUAES, University of Manchester, UK, 2013.**Co-PI of ICAR, NAHEP and World Bank funded project on Conservation Agriculture. So, far he has successfully guided 19 Ph D scholars and 77 M Sc students from BCKV CU, KU, Vidyasagar University. **He is the fellow, ISEE, IARI, New Delhi**

Phone: 91-9674419142; Home: B-15/36, Kalyani, Nadia;WB; 741235

Dr. Prasenjit Pal is presently working as Assistant Professor (SG) in Department of Fisheries Extension, Economics and Statistics at College of Fisheries, Central Agricultural University, Imphal. He has been in teaching, research and extension for the last nine years in the Department of Extension and Social Sciences, College of Fisheries, CAU (I). He obtained his M.Sc degree in Agricultural Statistics from Indian Agricultural Research Institute (IASRI), New Delhi, in 2006 and Ph.D degree in Agricultural Statistics from Bidhan Chandra Krishi Viswa Vidyalay (BCKVV) in 2011. He is associated with many professional organizations. He authored 70 research publications in reputed National (54) and International Journals (16), 3 Books and good number of popular articles/chapters/extension bulletin as on date. Besides this, he is associated with many administrative responsibilities. He has organized many training programmes, webinars. He is the life member of several scientific societies.

Research Methodology

— *Techniques and Methods* —

– *Authors* –

S K Acharya

P Pal

2023

Daya Publishing House®

A Division of

Astral International Pvt. Ltd.

New Delhi – 110002

ISBN: 9789359191744

Published by : **Daya Publishing House®**
A Division of
Astral International Pvt. Ltd.
– ISO 9001:2015 Certified Company –
4736/23, Ansari Road, Darya Ganj
New Delhi-110 002
Ph. 011-43549197, 8130496929
e-mail: info@astralint.com
Website: www.astralint.com

BIDHAN CHANDRA KRISHI VISWAVIDYALAYA
Office of the Dean, Post Graduate Studies

Prof. J.K. Hore
Dean,
Post Graduate Studies (Actg)

P.O.Krishi Viswavidyalaya, Mohanpur – 741252 Dist. Nadia, West Bengal.
From Kolkata-953473-278605: PBX :(STD-03473) 222269/70-Extn.57
Email deanpgsbckv1974@gmail.com, Fax – 03473 – 278605
Mobile 9477473506, 06292269108

No. Date- 12.07.2022

FOREWORD

Effective and successful research work cannot be executed without proper design, methodology and application. Higher education in India and elsewhere in other parts of the world needs a strong foundation on research methodology to foster the generation of innovation and excellence in science and education. In several cases, failure of research or generation of unsatisfactory results happens due to inadequate research methodology.

The book being published from Astral International on Research Methodology: Techniques and Methods authored by Prof. S. K. Acharya and Dr. P. Pal, to my mind, has been a time felt and extremely useful publication to serve the scholastic and analytical cultivation of post graduate students, faculties across the disciplines, and management professionals especially who are engaged in impact and output analysis.

The book would be much effective to guide and drive the young researchers and faculties how to go for successful thesis writing and analysis through different steps. I personally know Prof. Acharya as a prolific writer and thinker and with no exception, the book will be enriched with his profile of teaching and research experiences over 33 years at Bidhan Chandra Krishi Viswavidyalaya Involvement of Dr P. Pal with the back ground of Agricultural Statistics from Central Agriculture University, Manipur, in co-authoring the book has added a new flavour to methodological thinking in research arena.

Hore
(J. K. Hore)

Preface

Quality and effective research needs proper research design. Without proper design, no research process can generate desired results. The basics of research design, tools and techniques are common to any research in any discipline. Without correct sampling, it's not possible to collect representative and reliable data. For detecting functional determinants, we need to apply proper statistical tools. And equally, without solid, analyzed data, we can't draw any empirical inferences. The scholastic journey through data, determinants and decision merits a strong and immaculate research design along with befitting statistical and mathematical design. The book has gone deeper into the designing of research to high end non-linear programming; sampling to multivariate analytical tools, and includes participatory research and techniques. We are hugely optimistic that the book will cater a wide spectrum of post graduate students and Ph. D. scholars, candidates for ICAR/UGC NET, and also a genre of college and university faculties in multidisciplinary pursuits and scholastic practices.

S K Acharya

P Pal

Contents

— *Chapter 1* —

Research: The Meaning, Approaches and Typology

1.1 Research

Research in simple terms refers to search for knowledge. It is a scientific and systematic search for information on a particular topic or issue. It is also known as the art of scientific investigation. Several social scientists have defined research in different ways.

In the Encyclopaedia of Social Sciences, D. Slesinger and M. Stephension (1930) defined research as "the manipulation of things, concepts or symbols for the purpose of generalizing to extend, correct or verify knowledge, whether that knowledge aids in the construction of theory or in the practice of an art". According to Redman and Mory (1923), research is a "systematized effort to gain new knowledge". It is an academic activity and therefore the term should be used in a technical sense. According to Clifford Woody (Kothari, 1988), research comprises "defining and redefining problems, formulating hypotheses or suggested solutions; collecting, organizing and evaluating data; making deductions and reaching conclusions; and finally, carefully testing the conclusions to determine whether they fit the formulated hypotheses"

Thus, research is an original addition to the available knowledge, which contributes to its further advancement. It is an attempt to pursue truth through the methods of study, observation, comparison and experiment. In sum, research is the search for knowledge, using objective and systematic methods to find solution to a problem.

1.1.1 Objectives of Research

The objective of research is to find answers to the questions by applying scientific procedures. In other words, the main aim of research is to find out the truth which is hidden and has not yet been discovered. Although every research study has its own specific objectives, the research objectives may be broadly grouped as follows:

1. To gain familiarity with new insights into a phenomenon (*i.e.*, formulate research studies);
2. To accurately portray the characteristics of a particular individual, group, or a situation (*i.e.*, descriptive research studies);
3. To analyze the frequency with which something occurs (*i.e.*, diagnostic research studies);
4. To examine the hypothesis of a causal relationship between two variables (*i.e.*, hypothesis-testing research studies).

1.1.2 Research Methods versus Methodology

Research methods include all those techniques/methods that are adopted for conducting research. Thus, research techniques or methods are the methods that the researchers adopt for conducting the research studies. on the other hand, research methodology is the way in which research problems are solved systematically. It is a science of studying how research is conducted scientifically. Under it, the researcher acquaints himself/herself with the various steps generally adopted to study a research problem, along with the underlying logic behind them. Hence, it is not only important for the researcher to know the research techniques/methods, but also the scientific approach called methodology.

1.1.3 Research Approaches

There are two main approaches to research, namely quantitative approach and qualitative approach. The quantitative approach involves the collection of quantitative data, which are put to rigorous quantitative analysis in a formal and rigid manner. This approach further includes experimental, inferential, and simulation approaches to research. Meanwhile, the qualitative approach uses the method of subjective assessment of opinions, behaviour and attitudes. Research in such a situation is a function of the researcher's impressions and insights. The results generated by this type of research are either in non-quantitative form or in the form which cannot be put to rigorous quantitative analysis. Usually, this approach uses techniques like in depth interviews, focus group interviews, and projective techniques.

1.1.4 Types of Research

There are different types of research. The basic ones are as follows.

Descriptive Versus Analytical

Descriptive research consists of surveys and fact-finding enquiries of different types. The main objective of descriptive research is describing the state of affairs as it prevails at the time of study. The term 'ex post facto research' is quite often used for descriptive research studies in social sciences and business research. The most distinguishing feature of this method is that the researcher has no control over the variables here. He/she has to only report what is happening or what has happened. Majority of the ex post facto research projects are used for descriptive studies in which the researcher attempts to examine phenomena, such as the consumers' preferences, frequency of purchases, shopping, *etc.* Despite the inability of the researchers to control the variables, ex post facto studies may also comprise attempts by them to discover the causes of the selected problem. The methods of research adopted in conducting descriptive research are survey methods of all kinds, including correlation and comparative methods. Meanwhile, in the Analytical research, the researcher has to use the already available facts or information, and analyze them to make a critical evaluation of the subject.

Applied Versus Fundamental

Research can also be applied or fundamental in nature. An attempt to find a solution to an immediate problem encountered by a firm, an industry, a business organization, or the society is known as applied research. Researchers engaged in such researches aim at drawing certain conclusions confronting a concrete social or business problem. On the other hand, fundamental research mainly concerns generalizations and formulation of a theory. In other words, "Gathering knowledge for knowledge's sake is termed 'pure' or 'basic' research" (Young in Kothari, 1988). Researches relating to pure mathematics or concerning some natural phenomenon are instances of Fundamental Research. Likewise, studies focusing on human behaviour also fall under the category of fundamental research. Thus, while the principal objective of applied research is to find a solution to some pressing practical problem, the objective of basic research is to find information with a broad base of application and add to the already existing organized body of scientific knowledge.

Quantitative Versus Qualitative

Quantitative research relates to aspects that can be quantified or can be expressed in terms of quantity. It involves the measurement of quantity or amount. Various available statistical and econometric methods are adopted for analysis in such research. Which includes correlation, regressions and time series analysis *etc.*, On the other hand, Qualitative research is concerned with qualitative phenomena, or more specifically, the aspects related to or involving quality or kind. For example, an important type of qualitative research is 'Motivation Research', which investigates into the reasons for certain human behaviour. The main aim of this type of research is discovering the underlying motives and desires of human beings by using in-depth interviews.

The other techniques employed in such research are story completion tests, sentence completion tests, word association tests, and other similar projective methods. Qualitative research is particularly significant in the context of behavioural sciences, which aim at discovering the underlying motives of human behaviour. Such research helps to analyze the various factors that motivate human beings to behave in a certain manner, besides contributing to an understanding of what makes individuals like or dislike a particular thing. However, it is worth noting that conducting qualitative research in practice is considerably a difficult task. Hence, while undertaking such research, seeking guidance from experienced expert researchers is important

Conceptual Versus Empirical

The research related to some abstract idea or theory is known as Conceptual Research. Generally, philosophers and thinkers use it for developing new concepts or for reinterpreting the existing ones. Empirical Research, on the other hand, exclusively relies on the observation or experience with hardly any regard for theory and system. Such research is data based, which often comes up with conclusions that can be verified through experiments or observation. Empirical research is also known as experimental type of research, in which it is important to first collect the facts and their sources, and actively take steps to stimulate the production of desired information. In this type of research, the researcher first formulates a working hypothesis, and then gathers sufficient facts to prove or disprove the stated hypothesis. He/she formulates the experimental design, which according to him/her would manipulate the variables, so as to obtain the desired information. This type of research is thus characterized by the researcher's control over the variables under study. In simple term, empirical research is most appropriate when an attempt is made to prove that certain variables influence the other variables in some way. Therefore, the results obtained by using the experimental or empirical studies are considered to be the most powerful evidences for a given hypothesis.

Other Types of Research

The remaining types of research are variations of one or more of the aforementioned type of research. They vary in terms of the purpose of research, or the time required to complete it, or may be based on some 8 other similar factors.

On the basis of time, research may either be in the nature of one-time or longitudinal time series research. While the research is restricted to a single time-period in the former case, it is conducted over several time-periods in the latter case. Depending upon the environment in which the research is to be conducted, it can also be laboratory research or field-setting research, or simulation research, besides being diagnostic or clinical in nature. Under such research, in-depth approaches or case study method may be employed to analyze the basic causal relations. These studies usually undertake a detailed in-depth analysis of the causes of certain events of interest, and use very small samples and sharp data collection methods. The research may also be explanatory in nature. Formalized research studies consist of substantial structure and specific hypotheses to be verified. As regards to historical research, sources like historical documents, remains, *etc.* Are utilized to study past events or ideas. It also includes philosophy of persons and groups of the past or any remote point of time.

Research has also been classified into decision-oriented and conclusion-oriented categories. The decision-oriented research is always carried out as per the need of a decision maker and hence, the researcher has no freedom to conduct the research according to his/her own desires. On the other hand, in the case of Conclusion-oriented research, the researcher is free to choose the problem, redesign the enquiry as it progresses and even change conceptualization as he/she wishes to.

An operation research is a kind of decision-oriented research, where in scientific method is used in providing the departments, a quantitative basis for decision-making with respect to the activities under their purview.

1.1.5 Importance of Knowing How to Conduct Research

The importance of knowing how to conduct research is listed below:

i. The knowledge of research methodology provides training to new researchers and enables them to do research properly. It helps them to develop disciplined thinking or a 'bent of mind' to objectively observe the field;

ii. The knowledge of doing research inculcates the ability to evaluate and utilize the research findings with confidence;

iii. The knowledge of research methodology equips the researcher with the tools that help him/her to make the observations objectively; and

iv. The knowledge of methodology helps the research consumers to evaluate research and make rational decisions.

1.1.6 Qualities of a Researcher

It is important for a researcher to possess certain qualities to conduct research. First and foremost, he being a scientist should be firmly committed to the 'articles of faith' of the scientific methods of research. This implies that a researcher should be a social science person in the truest sense. Sir Michael Foster cited by (Wilkinson and Bhandarkar, 1979) identified a few distinct qualities of a scientist. According to him, a true research scientist should possess the following qualities:

1. First of all, the nature of a researcher must be of the temperament that vibrates in unison with the theme which he is searching. Hence, the seeker of knowledge must be truthful with truthfulness of nature, which is much more important, much more exacting than what is sometimes known as truthfulness. The truthfulness relates to the desire for accuracy of observation and precision of statement. Ensuring facts is the principle rule of science, which is not an easy matter. The difficulty may arise due to untrained eye, which fails to see anything beyond what it has the power of seeing and sometimes even less than that. This may also be due to the lack of discipline in the method of science. An unscientific individual often remains satisfied with the expressions like approximately, almost, or nearly, which is never what nature is. A real research cannot see two things which differ, however minutely, as the same.
2. A researcher must possess an alert mind. Nature is constantly changing and revealing itself through various ways. A scientific researcher must be keen and watchful to notice such changes, no matter how small or insignificant they may appear. Such receptivity has to be cultivated slowly and patiently over time by the researcher through practice. An individual who is ignorant or not alert and receptive during his research will not make a good researcher. He will fail as a good researcher if he has no keen eyes or mind to observe the unusual changes behind the routine. Research 10 demands a systematic immersion into the subject matter by the researcher grasp even the slightest hint that may culminate into significant research problems. In this context, Cohen and Negal cited by (Selltiz *et al.*, 1965; Wilkinson and Bhandarkar, 1979) state that "the ability to perceive in some brute experience the occasion of a problem is not a common talent among men... it is a mark of scientific genius to be sensitive to difficulties where less gifted people pass by untroubled by doubt".
3. Scientific enquiry is pre-eminently an intellectual effort. It requires the moral quality of courage, which reflects the courage of a steadfast endurance. The process of conducting research is not an easy task. There are occasions when a research scientist might feel defeated or completely lost. This is the stage when a researcher would need immense courage and the sense of conviction. The researcher must learn the art of enduring intellectual hardships. In the words of Darwin, "It's dogged that does it". In order to cultivate the afore-mentioned three qualities of a researcher, a fourth one may be added. This is the quality of making statements cautiously. According to Huxley, the assertion that outstrips the evidence is not only a blunder but a crime

(Thompson, 1975). A researcher should cultivate the habit of reserving judgment when the required data are insufficient.

1.1.7 Significance of Research

According to a famous Hudson Maxim, "All progress is born of inquiry. Doubt is often better than overconfidence, for it leads to inquiry, and inquiry leads to invention". It brings out the significance of research, increased amount of which makes the progress possible. Research encourages scientific and inductive thinking, besides promoting the development of logical habits of thinking and organization. The role of research in applied economics in the context of an economy or business is greatly increasing in modern times. The increasingly complex nature of government and business has raised the use of research in solving operational problems. Research assumes significant role in the formulation of economic policy for both, the government and business. It provides the basis for almost all government policies of an economic system. Government budget formulation, for example, depends particularly on the analysis of needs and desires of people, and the availability of revenues, which requires research.

Research helps to formulate alternative policies, in addition to examining the consequences of these alternatives. Thus, research also facilitates the decision-making of policy-makers, although in itself is not a part of research. In the process, research also helps in the proper allocation of a country's scarce resources. Research is also necessary for collecting information on the social and economic structure of an economy to understand the process of change occurring in the country. Collection of statistical information, though not a routine task, involves various research problems. Therefore, large staff of research technicians or experts is engaged by the government these days to undertake this work.

Thus, research as a tool of government economic policy formulation involves three distinct stages of operation:

1. **Stage 1:** Investigation of economic structure through continual compilation of facts;
2. **Stage 2:** Diagnosis of events that are taking place and analysis of the forces underlying them; and
3. **Stage 3:** The prognosis *i.e.,* the prediction of future developments (Wilkinson and Bhandarkar, 1979).

Research also assumes significance in solving various operational and planning problems associated with business and industry. In several ways, operations research, market research and motivational research are vital and their results assist in taking business decisions. Market research refers to the investigation of the structure and development of a market for the formulation of efficient policies relating to purchases, production and sales. Operational research relates to the application of logical, mathematical, and analytical techniques to find solution to business problems, such as cost minimization or profit maximization, or the optimization problems. Motivational research helps to determine why people behave in the manner they do with respect to market characteristics. More specifically, it is concerned with the analysis of the motivations underlying consumer behaviour. All these researches are very useful for business and industry, and are responsible for business decision-making. Research is equally important to social scientists for analyzing the social relationships and seeking explanations to various social problems. It gives intellectual satisfaction of knowing things for the sake of knowledge. It also possesses the practical utility for the social scientist to gain knowledge so as to be able to do something better or in a more efficient manner. The research in social sciences is concerned with both knowledge for its own sake, and knowledge for what it can contribute to solve practical problems.

1.1.8 Ethical Issues in Research

Research ethics involve requirements on daily work, the protection of dignity of subjects and the publication of the information in the research. Ethical issues, conflicting values, and ambiguity in decision making, are recurrently emerging from literature review on research. The major ethical issues in conducting research are: a) Informed consent, b) Beneficence- Do not harm c) Respect for anonymity and confidentiality d) Respect for privacy. There are a number of key phrases that describe the system of ethical protections that the contemporary social/biological/medical research establishments have created to try to protect better the rights of their research participants. The principle of ***voluntary participation*** requires that people not be coerced into participating in research. This is especially relevant where researchers had previously relied on 'captive audiences' for their subjects—prisons, universities, and places like that. Closely related to the notion of voluntary participation is the requirement of ***informed consent***. The prospective research participants must be fully informed about the procedures and risks involved in research and must give their consent to participate. Ethical standards also require that researchers not put participants in a situation where they might be at ***risk of harm*** as a result of their participation. Harm can be defined as both physical and psychological. There are two standards that are applied in order to help protect the privacy of research participants. Almost all research guarantees the participants' ***confidentiality*** — they are assured that identifying information will not be made available to anyone who is not directly involved in the study. The stricter standard is the principle of ***anonymity*** which essentially means that the participant will remain anonymous throughout the study-even to the researchers themselves. Clearly, the anonymity standard is a stronger guarantee of privacy, but it is sometimes difficult to accomplish, especially in situations where participants have to be measured at multiple time points (*e.g.*, a pre-post study). Increasingly, researchers have had to deal with the ethical issue of a person's ***right to service***. Good research practice often requires the use of a no-treatment control group — a group of participants who don't get the treatment or program that is being studied. But when that treatment or program may have beneficial effects, persons assigned to the no-treatment control may feel their rights to equal access to services are being curtailed.

1.2 Research Process

Research process consists of a series of steps or actions required for effectively conducting research. The following are the steps that provide useful procedural guidelines regarding the conduct of research: (1) Formulating the research problem; (2) Extensive literature survey; (3) Developing hypothesis; (4) Preparing the research design; (5) Determining sample design; (6) Collecting data; (7) Execution of the project; (8) Analysis of data; (9) Hypothesis testing; (10) Generalization and interpretation, and (11) Preparation of the report or presentation of the results. In other words, it involves the formal write-up of conclusions.

1.3 Research Problem

The first and foremost stage in the research process is to select and properly define the research problem. A researcher should first identify a problem and formulate it, so as to make it amenable or susceptible to research. In general, a research problem refers to an unanswered question that a researcher might encounter in the context of either a theoretical or practical situation, which he/she would like to answer or find a solution to. A research problem is generally said to exist if the following conditions emerge (Kothari, 1988): i. There should be an individual or an organization, say X, to whom the Problem can be attributed. The individual or the organization is situated in an environment Y, which is governed by certain uncontrolled variables Z; 13 ii. There should be at least two courses of action to be pursued, say A1 and A2. These courses of action are defined by one or more values of the controlled variables. For example, the number of items purchased at a specified time is said to be one course of action. iii. There should be at least two alternative possible outcomes of the said courses of action, say B1 and B2. Of them, one alternative should be preferable to the other. That is, at least one outcome should be what the researcher wants, which becomes an objective. iv. The courses of possible action available must offer a chance to the researcher to achieve the objective, but not the equal chance. Therefore, if P(Bj/X, A, Y) represents the probability of the occurrence of an outcome Bj when X selects Aj in Y, then $P(B1/X, A1, Y) \neq P(B1/X, A2, Y)$. Putting it in simple words, it means that the choices must not have equal efficiencies for the desired outcome. Above all these conditions, the individual or organization may be said to have arrived at the research problem only if X does not know what course of action to be taken is the best. In other words, X should have a doubt about the solution. Thus, an individual or a group of persons can be said to have a problem if they have more than one desired outcome. They should have two or more alternative courses of action, which have some but not equal efficiency. This is required for probing the desired objectives, such that they have doubts about the best course of action to be taken. Thus, the components of a research problem may be summarized as:

1. There should be an individual or a group who have some difficulty or problem.
2. There should be some objective(s) to be pursued. A person or an organization who wants nothing cannot have a problem.
3. There should be alternative ways of pursuing the objective the researcher wants to pursue. This implies that there should be more than one alternative means available to the researcher. This is because if the researcher has no choice of alternative means, he/she would not have a problem.
4. There should be some doubt in the mind of the researcher about the choice of alternative means. This implies that research should answer the question relating to the relative efficiency or suitability of the possible alternatives.
5. There should be a context to which the difficulty relates. Thus, identification of a research problem is the pre-condition to conducting research.

A research problem is said to be the one which requires a researcher to find the best available solution to the given problem. That is, the researcher needs to find out the best course of action through which the research objective may be achieved optimally in the context of a given situation. Several factors may contribute to making the problem complicated. For example, the environment may alter, thus affecting the efficiencies of the alternative courses of action taken or the quality of the outcomes. The number of alternative courses of action might be very large and the individual not involved in making the decision may be affected by the change in environment and may react to it favourably or unfavourably. Other similar factors are also likely to cause such changes in the context of research, all of which may be considered from the point of view of a research problem.

1.4 Research Design

The most important step after defining the research problem is preparing the design of the research project, which is popularly known as the 'research design'. A research design helps to decide upon issues like what, when, where, how much, by what means *etc.* with regard to an enquiry or a research study. A research design is the arrangement of conditions for collection and analysis of data in a manner that aims to combine relevance to the research purpose with economy in procedure. In fact, research design is the conceptual structure within which research is conducted; it constitutes the blueprint for the collection, measurement and analysis of data (Selltiz *et al.*, 1962). Thus, research design provides an outline of what the researcher is going to do in terms of framing the hypothesis, its operational implications and the final data analysis. Specifically, the research design highlights decisions which include:

1. The nature of the study
2. The purpose of the study
3. The location where the required data can be collected
4. What time period the study would cover
5. The type of sample design that would be used
6. The techniques of data collection that would be used
7. The methods of data analysis that would be adopted and
8. The manner in which the report would be prepared.

In view of the stated research design decisions, the overall research design may be divided into the following (Kothari 1988):

a. The sampling design that deals with the method of selecting items to be observed for the selected study;
b. The observational design that relates the conditions under which the observations are to be made;
c. The statistical design that concerns with the question of how many items are to be observed, and how the information and data gathered are to be analysed; and,
d. The operational design that deals with the techniques by which the procedures specified in the sampling, statistical and observational designs can be carried out.

1.4.1 Features of Research Design

The important features of Research Design may be outlined as follows: i. It constitutes a plan that identifies the types and sources of information required for the research problem; ii. It constitutes a strategy that specifies the methods of data collection and analysis which would be adopted; and iii. It also specifies the time period of research and monetary budget involved in conducting the study, which comprise the two major constraints of undertaking any research.

Concepts Relating to Research Design

Some of the important concepts relating to Research Design are discussed below:

1. *Dependent and Independent Variables*

A magnitude that varies is known as a variable. The concept may assume different quantitative values like height, weight, income *etc.* Qualitative variables are not quantifiable in the strictest sense of the term. However, the qualitative phenomena may also be quantified in terms of the presence or absence of the attribute(s) considered. The phenomena that assume different values quantitatively even in decimal points are known as 'continuous variables'. But all variables need not be continuous. Values that can be expressed only in integer values are called 'non-continuous variables'. In statistical terms, they are also known as 'discrete variables'. For example, age is a continuous variable, whereas the number of children is a non-continuous variable.

When changes in one variable depend upon the changes in other variable or variables, it is known as a dependent or endogenous variable, and the variables that cause the changes in the dependent variable are known as the independent or explanatory or exogenous variables. For example, if demand depends upon price, then demand is a dependent variable, while price is the independent variable. And, if more variables determine demand, like income and price of the substitute commodity, then demand also depends upon them in addition to the price of original commodity. In other words, demand is a dependent variable which is determined by the independent variables like price of the original commodity, income and price of substitutes

Extraneous Variables

The independent variables which are not directly related to the purpose of the study but affect the dependent variables, are known as extraneous variables. For instance, assume that a researcher wants to test the hypothesis that there is a relationship between children's school performance and their self-confidence, in which case the latter is an independent variable and the former, a dependent variable. In this context, intelligence may also influence the school performance. However, since it is not directly related to the purpose of the study undertaken by the 17 researcher, it would be known as an extraneous variable.

2. *The Influence Caused by the Extraneous Variable(s) on the Dependent Variable*

It is technically called the 'experimental error'. Therefore, a research study should always be framed in such a manner that the influence of extraneous variables on the dependent variable/s is completely controlled, and the influence of independent variable/s is clearly evident.

3. *Control*

One of the most important features of a good research design is to minimize the effect of extraneous variable(s). Technically, the term 'control' is used when a researcher designs the study in such a manner that it minimizes the effects of extraneous variables. The term 'control' is used in experimental research to reflect the restrain in experimental conditions

4. *Confounded Relationship*

The relationship between the dependent and independent variables is said to be confounded by an extraneous variable, when the dependent variable is not free from its effects.

5. *Research Hypothesis*

When a prediction or a hypothesized relationship is tested by adopting scientific methods, it is known as research hypothesis. The research hypothesis is a predictive statement which relates to a dependent variable and an independent variable. Generally, a research hypothesis must consist of at least one dependent variable and one independent variable. Whereas, the relationships that are assumed but not to be tested are predictive statements that are not to be objectively verified, thus are not classified as research hypotheses. 6. Experimental and Non-experimentation

6. *Hypothesis Testing Research*

When the objective of a research is to test a research hypothesis, it is known as hypothesis-testing research. Such research may be in the nature of experimental design or non-experimental design. The research in which the independent variable is manipulated is known as 'experimental hypothesis-testing research', whereas the research in which the independent 18 variable is not manipulated is termed as 'non-experimental hypothesis testing research'. For example, assume that a researcher wants to examine whether family income influences the school attendance of a group of students, by calculating the coefficient of correlation between the two variables. Such an example is known as a non-experimental hypothesis testing research, because the independent variable - family income is not manipulated here. Again assume that the researcher randomly selects 150 students from a group of students who pay their school fees regularly and then classifies them into two sub-groups by randomly including 75 in Group A, whose parents have regular earning, and 75 in Group B, whose parents do not have regular earning. Assume that at the end of the study, the researcher conducts a test on each group in order to examine the effects of regular earnings of the parents on the school attendance of the student. Such a study is an example of experimental hypothesis-testing research, because in this particular study the independent variable regular earnings of the parents have been manipulated.

7. *Experimental and Control Groups*

When a group is exposed to usual conditions in an experimental hypothesis-testing research, it is known as 'control group'. On the other hand, when the group is exposed to certain new or special condition, it is known as an 'experimental group'. In the afore-mentioned example, Group A can be called as control group and Group B as experimental group. If both the groups, A and B are exposed to some special feature, then both the groups may be called as 'experimental groups'. A research design may include only the experimental group or both the experimental and control groups together.

8. *Treatments*

Treatments refer to the different conditions to which the experimental and control groups are subject to. In the example considered, the two treatments are the parents with regular earnings and those with no regular earnings. Likewise, if a research study attempts to examine through an experiment the comparative effect of three different types of fertilizers on the yield of rice crop, then the three types of fertilizers would be treated as the three treatments.

9. *Experiment*

Experiment refers to the process of verifying the truth of a statistical hypothesis relating to a given research problem. For instance, an experiment may be conducted to examine the yield of a certain new variety of rice crop developed. Further, Experiments may be categorized into two types, namely, 'absolute experiment' and 'comparative experiment'. If a researcher wishes to determine the impact of a chemical fertilizer on the yield of a particular variety of rice crop, then it is known as absolute experiment. Meanwhile, if the researcher wishes to determine the impact of chemical fertilizer as compared to the impact of bio-fertilizer, then the experiment is known as a comparative experiment. 10. Experimental Unit(s): Experimental units refer to the pre-determined plots, characteristics or the blocks, to which different treatments are applied. It is worth mentioning here that such experimental units must be selected with great caution.

1.4.2 Types of Research Design

There are different types of research designs. They may be broadly categorized as:

1. Exploratory Research Design;
2. Descriptive and Diagnostic Research Design; and
3. Hypothesis-Testing Research Design.
4. Ex-Post-Facto Research Design

1. Exploratory Research Design

The Exploratory Research Design is known as formulate research design. The main objective of using such a research design is to formulate a research problem for an in-depth or more precise investigation, or for developing a working hypothesis from an operational aspect. The major purpose of such studies is the discovery of ideas and insights. Therefore, such a research design suitable for such a study should be flexible enough to provide opportunity for considering different dimensions of the problem under study. The in-built flexibility in research design is required as the 20 initial research problem would be transformed into a more precise one in the exploratory study, which in turn may necessitate changes in the research procedure for collecting relevant data.

Usually, the following three methods are considered in the context of a research design for such studies. They are:

(a) Survey of related literature;

(b) Experience survey; and

(c) Analysis of 'insight-stimulating' instances.

2. Descriptive and Diagnostic Research Design

A Descriptive Research Design is concerned with describing the characteristics of a particular individual or a group. Meanwhile, a diagnostic research design determines the frequency with which a variable occurs or its relationship with another variable. In other words, the study analyzing whether a certain variable is associated with another comprises a diagnostic research study. On the other hand, a study that is concerned with specific predictions or with the narration of facts and characteristics related to an individual, group or situation, are instances of descriptive research studies. Generally, most of the social research design falls under this category. As a research design, both the descriptive and diagnostic studies share common requirements, hence they are grouped together. However, the procedure to be used and the research design need to plan carefully. The research design must also make appropriate provision for protection against bias and thus maximize reliability, with due regard to the completion of the research study in an economical manner. The research design in such studies should be rigid and not flexible.

Besides, it must also focus attention on the following:

a) Formulation of the objectives of the study, b) Proper designing of the methods of data collection, c) Sample selection, d) Data collection, e) Processing and analysis of the collected data, and f) Reporting the findings.

3. Hypothesis-Testing Research Design

Hypothesis-Testing Research Designs are those in which the researcher tests the hypothesis of causal relationship between two or more variables. These studies require procedures that would not only decrease bias and enhance reliability, but also facilitate deriving inferences about the causality. Generally, experiments satisfy such requirements. Hence, when research design is discussed in such studies, it often refers to the design of experiments.

4. Ex-Post-Facto Research Designs

An ex post facto research design is a method in which groups with qualities that already exist are compared on some dependent variable. Also known as "after the fact" research, an ex post facto design is considered quasi-experimental because the subjects are not randomly assigned - they are grouped based on a particular characteristic or trait.

Although differing groups are analyzed and compared in regards to independent and dependent variables it is not a true experiment because it lacks random assignment. The assignment of subjects to different groups is based on whichever variable is of interest to the researchers. For example, a researcher is interested in how weight influences self-esteem levels in adults. So the participants would be separated into differing groups (underweight, normal weight, overweight) and their self esteem levels measured. This is an ex post facto design because a pre-existing characteristic (weight) was used to form the groups.

1.4.3 Importance of Research Design

The need for a research design arises out of the fact that it facilitates the smooth conduct of the various stages of research. It contributes to making research as efficient as possible, thus yielding the maximum information with minimum effort, time and expenditure. A research design helps to plan in advance, the methods to be employed for collecting the relevant data and the techniques to be adopted for their analysis. This would help in pursuing the objectives of the research in the best possible manner, provided the available staff, time and money are given. Hence, the research design should be prepared with utmost care, so as to avoid any error that may disturb the entire project. Thus, research design plays a crucial role in attaining the reliability of the results obtained, which forms the strong foundation of the entire process of the research work.

1.4.4 Characteristics of a Good Research Design

A good research design often possesses the qualities of being flexible, suitable, efficient, and economical and so on. Generally, a research design which minimizes bias and maximizes the reliability of the data collected and analyzed is considered a good design (Kothari 1988). A research design which does not allow even the smallest experimental error is said to be the best design for investigation.

Further, a research design that yields maximum information and provides an opportunity of viewing the various dimensions of a research problem is considered to be the most appropriate and efficient design. Thus, the question of a good design relates to the purpose or objective and nature of the research problem studied. While a research design may be good, it may not be equally suitable to all studies.

In other words, it may be lacking in one aspect or the other in the case of some other research problems. Therefore, no single research design can be applied to all types of research problems. A research design suitable for a specific research problem would usually involve the following considerations:

i. The methods of gathering the information;
ii. The skills and availability of the researcher and his/her staff, if any;
iii. The objectives of the research problem being studied;
iv. The nature of the research problem being studied; and v. The available monetary support and duration of time for the research work.

1.5 Case Study Research

The method of exploring and analyzing the life or functioning of a social or economic unit, such as a person, a family, a community, an institution, a firm or an industry is called case study method. The objective of case study method is to examine the factors that cause the behavioural patterns of a given unit and its relationship with the environment.

The data for a study are always gathered with the purpose of tracing the natural history of a social or economic unit, and its relationship with the social or economic factors, besides the forces involved in its environment. Thus, a researcher conducting a study using the case study method attempts to understand the complexity of factors that are operative within a social or economic unit as an integrated totality.

Burgess (Kothari, 1988) described the special significance of the case study in understanding the complex behaviour and situations in specific detail. In the context of social research, he called such data as social microscope

1.5.1 Criteria for Evaluating Adequacy of Case Study

John Dollard (Dollard, 1935) specified seven criteria for evaluating the adequacy of a case or life history in the context of social research. They are:

i. The subject being studied must be viewed as a specimen in a cultural set up. That is, the case selected from its total context for the purpose of study should be considered a member of the particular cultural group or community. The scrutiny of the life history of the individual must be carried out with a view to identify the community values, standards and shared ways of life.
ii. The organic motors of action should be socially relevant. This is to say that the action of the individual cases should be viewed as a series of reactions to social stimuli or situations. To put in simple words, the social meaning of behaviour should be taken into consideration.
iii. The crucial role of the family-group in transmitting the culture should be recognized. This means, as an individual is the member of a family, the role of the family in shaping his/her behavior should never be ignored.
iv. The specific method of conversion of organic material into social behaviour should be clearly demonstrated. For instance, case-histories

that discuss in detail how basically a biological organism, that is man, gradually transforms into a social person are particularly important.

v. The constant transformation of character of experience from childhood to adulthood should be emphasized. That is, the life-history should portray the inter-relationship between the individual's various experiences during his/her life span. Such a study provides a comprehensive understanding of an individual's life as a continuum.

vi. The 'social situation' that contributed to the individual's gradual transformation should carefully and continuously be specified as a factor. One of the crucial criteria for life-history is that an individual's life should be depicted as evolving itself in the context of a specific social situation and partially caused by it.

vii. The life-history details themselves should be organized according to some conceptual framework, which in turn would facilitate their generalizations at higher levels.

These criteria discussed by Dollard emphasize the specific link of co-ordinated, related, continuous and configured experience in a cultural pattern that motivated the social and personal behavior. Although, the criteria indicated by Dollard are principally perfect, some of them are difficult to put to practice.

Dollard (1935) attempted to express the diverse events depicted in the life-histories of persons during the course of repeated interviews by utilizing psycho-analytical techniques in a given situational context. His criteria of life-history originated directly from this experience. While the life-histories possess independent significance as research documents, the interviews recorded by the investigators can afford, as Dollard observed, "rich insights into the nature of the social situations experienced by them".

It is a well-known fact that an individual's life is very complex. Till date there is hardly any technique that can establish some kind of uniformity, and as a result ensure the cumulative of case-history materials by isolating the complex totality of a human life. Nevertheless, although case history data are difficult to put to rigorous analysis, a skillful handling and interpretation of such data could help in developing insights into cultural conflicts and problems arising out of cultural-change.

Gordon Allport in (Kothari 1988) has recommended the following aspects so as to broaden the perspective of case-study data:

i. If the life-history is written in first person, it should be as comprehensive and coherent as possible.

ii. Life-histories must be written for knowledgeable persons, that is, if the enquiry of study is sociological in nature, the researcher should 25 write it on the assumption that it would be read largely by sociologists only.

iii. It would be advisable to supplement case study data by observational, statistical and historical data, as they provide standards for assessing the reliability and consistency of the case study materials. Further, such data offer a basis for generalizations.

iv. Efforts must be made to verify the reliability of life-history data by examining the internal consistency of the collected material, and by repeating the interviews with the concerned person. Besides this, personal interviews with the persons who are well-acquainted with him/her, belonging to his/her own group should be conducted.

v. A judicious combination of different techniques for data-collection is crucial for collecting data that are culturally meaningful and scientifically significant.

vi. Life-histories or case-histories may be considered as an adequate basis for generalization to the extent that they are typical or representative of a certain group.

vii. The researcher engaged in the collection of case study data should never ignore the unique or typical cases. He/she should include them as exceptional cases.

Case histories are filled with valuable information of a personal or private nature. Such information not only helps the researcher to portray the personality of the individual, but also the social background that contributed to it. Besides, it also helps in the formulation of relevant hypotheses. In general, although Blummer (in Wilkinson and Bhandarkar, 1979) was critical of documentary material, he gave due credit to case histories by acknowledging the fact that the personal documents offer an opportunity to the researcher to develop his/her spirit of enquiry. The analysis of a particular subject would be more effective if the researcher acquires close acquaintance with it through personal documents. However, Blummer also acknowledges the limitations of the personal documents. According to him, such documents do not entirely fulfil the criteria of adequacy, reliability, and representativeness. Despite these shortcomings, avoiding their use in any scientific study of personal life would be wrong, as these documents become necessary and significant for both theory-building and practice. In spite of these formidable limitations, case study data are used by anthropologists, sociologists, economists and industrial psychiatrists. Gordon Allport (Kothari, 1988) strongly recommends the use of case study data for in-depth analysis of a subject. For, it is one's acquaintance with an individual that creates a desire to know his/her nature and understand them. The first stage involves understanding the individual and all the complexity of his/her nature. As a consequence, the important emotional organizations, anchorages and natural identifications characterizing the personal life of the individual might not yield adequate representation. Hence, the researcher should understand the life of the subject. Therefore, the totality of life-processes reflected in the well-ordered life-history documents become invaluable source of stimulating insights. Such life-history documents provide the basis for comparisons that contribute to statistical generalizations and help to draw inferences regarding the uniformities in human behaviour, which are of great value.

1.6 Hypothesis

"Hypothesis may be defined as a proposition or a set of propositions set forth as an explanation for the occurrence of some specified group of phenomena either asserted merely as a provisional conjecture to guide some investigation in the light of established facts" (Kothari, 1988). A research hypothesis is quite often a predictive statement, which is capable of being tested using scientific methods that involve an independent and some dependent variables. For instance, the following statements may be considered:

i. "Students who take tuitions perform better than the others who do not receive tuitions" or,
ii. "The female students perform as well as the male student

These two statements are hypotheses that can be objectively verified and tested. Thus, they indicate that a hypothesis states what one is looking for. Besides, it is a proposition that can be put to test in order to examine its validity. A hypothesis should have the following characteristic features.

i. A hypothesis must be precise and clear. If it is not precise and clear, then the inferences drawn on its basis would not be reliable.
ii. A hypothesis must be capable of being put to test. Quite often, the research programmes fail owing to its incapability of being subject to testing for validity. Therefore, some prior study may be conducted by the researcher in order to make a hypothesis testable. A hypothesis "is tested if other deductions can be made from it, which in turn can be confirmed or disproved by observation" (Kothari, 1988).
iii. A hypothesis must state relationship between two variables, in the case of relational hypotheses.
iv. A hypothesis must be specific and limited in scope. This is because a simpler hypothesis generally would be easier to test for the researcher. And therefore, he/she must formulate such hypotheses.
v. As far as possible, a hypothesis must be stated in the simplest language, so as to make it understood by all concerned. However, it should be noted that simplicity of a hypothesis is not related to its significance.
vi. A hypothesis must be consistent and derived from the most known facts. In other words, it should be consistent with a substantial body of established facts. That is, it must be in the form of a statement which is most likely to occur.
vii. A hypothesis must be amenable to testing within a stipulated or reasonable period of time. No matter how excellent a hypothesis, a

researcher should not use it if it cannot be tested within a given period of time, as no one can afford to spend a life-time on collecting data to test it.

viii. A hypothesis should state the facts that give rise to the necessity of looking for an explanation. This is to say that by using the hypothesis, and other known and accepted generalizations, a researcher must be able to derive the original problem condition. Therefore, a hypothesis should explain what it actually wants to explain, and for this it should also have an empirical reference.

1.6.1 Concepts Relating to Testing of Hypotheses

Testing of hypotheses requires a researcher to be familiar with various concepts concerned with it such as:

1. Null Hypothesis and Alternative Hypothesis

In the context of statistical analysis, hypotheses are of two types *viz.*, null hypothesis and alternative hypothesis. When two methods A and B are compared on their relative superiority, and it is assumed that both the methods are equally good, then such a statement is called as the null hypothesis. On the other hand, if method A is considered relatively superior to method B, or vice-versa, then such a statement is known as an alternative hypothesis. The null hypothesis is expressed as **H0**, while the alternative hypothesis is expressed as Ha. For example, if a researcher wants to test the hypothesis that the population mean (μ) is equal to the hypothesized mean (H0) = 100, then the null hypothesis should be stated as the population mean is equal to the hypothesized mean 100. Symbolically it may be written as:-

H0: = μ = μ H0 = 100

If sample results do not support this null hypothesis, then it should be concluded that something else is true. The conclusion of rejecting the null hypothesis is called as alternative hypothesis H1. To put it in simple words, the set of alternatives to the null hypothesis is termed as the alternative hypothesis. If H0 is accepted, then it implies that Ha is being rejected. On the other hand, if H0 is rejected, it means that Ha is being accepted. For H0: μ = μ H0 = 100, the following three possible alternative hypotheses may be considered:

Alternative Hypothesis	*To be Read as follows*
H1: μ ≠ μ H0	The alternative hypothesis is that the population mean is not equal to 100, *i.e.*, it could be greater than or less than 100
H1: μ > μ H0	The alternative hypothesis is that the population mean is greater than 100
H1: μ < μ H0	The alternative hypothesis is that the population mean is less than 100

2. Test of Hypothesis or Decision Rule

Suppose the given hypothesis is H0 and the alternative hypothesis H1, then the researcher has to make a rule known as the decision rule. According to the decision rule, the researcher accepts or rejects H0. For example, if the H0 is that certain students are good against the H1 that all the students are good, then the researcher should decide the number of items to be tested and the criteria on the basis of which to accept or reject the hypothesis.

3. Type I and Type II Errors

As regards the testing of hypotheses, a researcher can make basically two types of errors. He/she may reject H0 when it is true, or accept H0 when it is not true. The former is called as Type I error and the latter is known as Type II error. In other words, Type I error implies the rejection of a hypothesis when it must have been accepted, while Type II error implies the acceptance of a hypothesis which must have been rejected. Type I error is denoted by α (alpha) and is known as α error, while Type II error is usually denoted by β (beta) and is known as β error.

4. One-Tailed and Two-Tailed Tests

These two types of tests are very important in the context of hypothesis testing. A two-tailed test rejects the null hypothesis, when the sample mean is significantly greater or lower than the hypothesized value of the mean of the population. Such a test is suitable when the null hypothesis is some specified value; the alternative hypothesis is a value that is not equal to the specified value of the null hypothesis.

1.6.2 Procedure of Hypothesis Testing

Testing a hypothesis refers to verifying whether the hypothesis is valid or not. Hypothesis testing attempts to check whether to accept or not to accept the null hypothesis. The procedure of hypothesis testing includes all the steps that a researcher undertakes for making a choice between the two alternative actions of rejecting or accepting a null hypothesis. The various steps involved in hypothesis testing are as follows:

1. Making a Formal Statement

This step involves making a formal statement of the null hypothesis (H0) and the alternative hypothesis (Ha). This implies that the hypotheses should be clearly stated within the purview of the research problem. For example, suppose a school teacher wants to test the understanding capacity of the students which must be rated more than 90 per cent in terms of marks, the hypotheses may be stated as follows:

Null Hypothesis H0: = 100 Alternative Hypothesis H1: > 100

2. Selecting a Significance Level

The hypotheses should be tested on a pre-determined level of significance, which should be specified. Usually, either 5 per cent level or 1 per cent level is considered for the purpose. The factors that determine the levels of significance are: (a) the magnitude of difference between the sample means; (b) the sample size: (c) the variability of measurements within samples; and (d) whether the hypothesis is directional or non-directional (Kothari, 1988). In sum, the level of significance should be sufficient in the context of the nature and purpose of enquiry.

3. Deciding the Distribution to Use

After making decision on the level of significance for hypothesis testing, the researcher has to next determine the appropriate sampling distribution. The choice to be made generally relates to normal distribution and the t-distribution. The rules governing the selection of the correct distribution are similar to the ones already discussed with respect to estimation.

4. Selection of a Random Sample and Computing an Appropriate Value

5. Calculation of the Probability

The next step for the researcher is to calculate the probability that the sample result would diverge as far as it can from expectations, under the situation when the null hypothesis is actually true.

6. Comparing the Probability

Another step involved consists of making a comparison of the probability calculated with the specified value of α, *i.e.* the significance level. If the calculated probability works out to be equal to or smaller than the value in case of one-tailed test, then the null hypothesis is to be rejected. On the other hand, if the calculated probability is greater, then the null hypothesis is to be accepted. In case the null hypothesis H0 is rejected, the researcher runs the risk of committing the Type I error. But, if the null hypothesis H0 is accepted, then it involves some risk (which cannot be specified in size as long as H0 is vague and not specific) of committing the Type II error.

1.7 Levels of Measurement

Before we can conduct a statistical analysis, we need to measure our dependent variable. Exactly how the measurement is carried out depends on the type of variable involved in the analysis. Different types are measured differently. To measure the time taken to respond to a stimulus, you might use a stop watch. Stop watches are of no use, of course, when it comes to measuring someone's attitude towards a political candidate. A rating scale is more appropriate in this case (with labels like "very favorable," "somewhat favorable," *etc.*). For a dependent variable such as "favorite color," you can simply note the color-word (like "red") that the subject offers. Although procedures for measurement differ in many ways, they can be classified using a few fundamental categories. In a given category, all of the procedures share some properties that are important for you to know about. The categories are called "scale types," or just "scales," and are described in this section. A higher level of measurement provides a better opportunity for statistical analysis and interpretation of data.

Nominal Scales

In nominal level numbers are assigned to objects or events which can be placed into mutually exclusive and exhaustive categories. When measuring using a nominal scale, one simply names or categorizes responses. Gender, handedness, favorite color, and religion are examples of variables measured on a nominal scale. The essential point about nominal scales is that they do not imply any ordering among the responses. For example, when classifying people according to their favorite color, there is no sense in which green is placed "ahead of" blue. Responses are merely categorized. Nominal scales embody the lowest level of measurement.

Ordinal Scales

In ordinal level numbers are assigned to objects or events which can be placed into mutually exclusive and exhaustive categories and be ordered into a greater or less than scale. A researcher wishing to measure consumers' satisfaction with their microwave ovens might ask them to specify their feelings as either "very dissatisfied," "somewhat dissatisfied," "somewhat satisfied," or "very satisfied." The items in this scale are ordered, ranging from least to most satisfied. This is what distinguishes ordinal from nominal scales. Unlike nominal scales, ordinal scales allow comparisons of the degree to which two subjects possess the dependent variable. For example, our satisfaction ordering makes it meaningful to assert that one person is more satisfied than another with their microwave ovens. Such an assertion reflects the first person's use of a verbal label that comes later in the list than the label chosen by the second person.

On the other hand, ordinal scales fail to capture important information that will be present in the other scales we examine. In particular, the difference between two levels of an ordinal scale cannot be assumed to be the same as the difference between two other levels. In our satisfaction scale, for example, the difference between the responses "very dissatisfied" and "somewhat dissatisfied" is probably not equivalent to the difference between "somewhat dissatisfied" and "somewhat satisfied." Nothing in our measurement procedure allows us to determine whether the two differences reflect the same difference in psychological satisfaction. Statisticians express this point by saying that the differences between adjacent scale values do not necessarily represent equal intervals on the underlying scale giving rise to the measurements. (In our case, the underlying scale is the true feeling of satisfaction, which we are trying to measure.)

What if the researcher had measured satisfaction by asking consumers to indicate their level of satisfaction by choosing a number from one to four? Would the difference between the responses of one and two necessarily reflect the same difference in satisfaction as the difference between the responses two and three? The answer is No. Changing the response format to numbers does not change the meaning of the scale. We still are in no position to assert that the mental step from 1 to 2 (for example) is the same as the mental step from 3 to 4.

Interval Scales

In interval level numbers are assigned to objects or events which can be categorized, ordered and assumed to have an equal distance between scale values. Interval scales are numerical scales in which intervals have the same interpretation throughout. As an example, consider the Fahrenheit scale of temperature. The difference between 30 degrees and 40 degrees represents the same temperature difference as the difference between 80 degrees and 90 degrees. This is because each 10-degree interval has the same physical meaning (in terms of the kinetic energy of molecules).

Interval scales are not perfect, however. In particular, they do not have a true zero point even if one of the scaled values happens to carry the name "zero." The Fahrenheit scale illustrates the issue. Zero degrees Fahrenheit does not represent the complete absence of temperature (the absence of any molecular kinetic energy). In reality, the label "zero" is applied to its temperature for quite accidental reasons connected to the history of temperature measurement. Since an interval scale has no true zero point, it does not make sense to compute ratios of temperatures. For example, there is no sense in which the ratio of 40 to 20 degrees Fahrenheit is the same as the ratio of 100 to 50 degrees; no interesting physical property is preserved across the two ratios. After all, if the "zero" label were applied at the temperature that Fahrenheit happens to label as 10 degrees, the two ratios would instead be 30 to 10 and 90 to 40, no longer the same! For this reason, it does not make sense to say that 80 degrees is "twice as hot" as 40 degrees. Such a claim would depend on an arbitrary decision about where to "start" the temperature scale, namely, what temperature to call zero (whereas the claim is intended to make a more fundamental assertion about the underlying physical reality).

Ratio Scales

In interval level numbers are assigned to objects or events which can be categorized, ordered and assumed to have an equal distance between scale values and have a real zero point. The ratio scale of measurement is the most informative scale and it is the highest level of measurement as it has all the properties of nominal, ordinal, interval scales, plus an absolute zero point. It is an interval scale with the additional property that its zero position indicates the absence of the quantity being measured. You can think of a ratio scale as the three earlier scales rolled up in one. Like a nominal scale, it provides a name or category for each object (the numbers serve as labels). Like an ordinal scale, the objects are ordered (in terms of the ordering of the numbers). Like an interval scale, the same difference at two places on the scale has the same meaning. And in addition, the same ratio at two places on the scale also carries the same meaning.

The Fahrenheit scale for temperature has an arbitrary zero point and is therefore not a ratio scale. However, zero on the Kelvin scale is absolute zero. This makes the Kelvin scale a ratio scale. For example, if one temperature is twice as high as another as measured on the Kelvin scale, then it has twice the kinetic energy of the other temperature.

Level/Scale	*Central Tendency*
Nominal	Mode
Ordinal	Median
Interval	Mean deviation
Ratio	Geometric mean, Coefficient of Variation

1.8 Assessment of Reliability and Validity

Assessing the reliability of study findings requires researchers to make judgments about the 'soundness' of the research in relation to the application and appropriateness of the methods undertaken and the integrity of the final conclusions. Reliability refers to the consistency of a measure. Researchers consider three types of consistency: over time (test-retest reliability), across items (internal consistency), and across different researchers (inter-rater reliability).

1. Test-Retest Reliability

When researchers measure a construct that they assume to be consistent across time, then the scores they obtain should also be consistent across time. Test-retest reliability is the extent to which this is actually the case. For example, intelligence is generally thought to be consistent across time. A person who is highly intelligent today will be highly intelligent next week. This means that any good measure of intelligence should produce roughly the same scores for this individual next week as it does today. Clearly, a measure that produces highly inconsistent scores over time cannot be a very good measure of a construct that is supposed to be consistent. Assessing test-retest reliability requires using the measure on a group of people at one time, using it again on the *same* group of people at a later time, and then looking at test-retest correlation between the two sets of scores. This is typically done by graphing the data in a scatter plot and computing Pearson's r. In general, a test-retest correlation of +0.80 or greater is considered to indicate good reliability. Again, high test-retest correlations make sense when the construct being measured is assumed to be consistent over time, which is the case for intelligence, self-esteem, and the Big Five personality dimensions. But other constructs are not assumed to be stable over time. The very nature of mood, for example, is that it changes. So a measure of mood that produced a low test-retest correlation over a period of a month would not be a cause for concern.

2. Internal Consistency

A second kind of reliability is internal consistency, which is the consistency of people's responses across the items on a multiple-item measure. In general, all the items on such measures are supposed to reflect the same underlying construct, so people's scores on those items should be correlated with each other. On the Rosenberg Self-Esteem Scale, people who agree that they are a person of worth should tend to agree that that they have a number of good qualities. If people's responses to the different items are not correlated with each other, then it would no longer make sense to claim that they are all measuring the same underlying construct. This is as true for behavioural and physiological measures as for self-report measures. For example, people might make a series of bets in a simulated game of roulette as a measure of their level of risk seeking. This measure would be internally consistent to the extent that individual participants' bets were consistently high or low across trials. Like test-retest reliability, internal consistency can only be assessed by collecting and analyzing data. One approach is to look at a split-half correlation. This involves splitting the items into two sets, such as the first and second halves of the items or the even- and odd-numbered items. Then a score is computed for each set of items, and the relationship between the two sets of scores is examined. A split-half correlation of +0.80 or greater is generally considered good internal consistency. Perhaps the most common measure of internal consistency used by researchers in psychology is a statistic called Cronbach's α (the Greek letter alpha). Conceptually, α is the mean of all possible split-half correlations for a set of items. For example, there are 252 ways to split a set of 10 items into two sets of five. Cronbach's α would be the mean of the 252 split-half correlations. Note that this is not how α is actually computed, but it is a correct way of interpreting the meaning of this statistic. Again, a value of 0+.80 or greater is generally taken to indicate good internal consistency.

3. Interrater Reliability

Many behavioural measures involve significant judgment on the part of an observer or a rater. Inter-rater reliability is the extent to which different observers are consistent in their judgments. For example, if you were interested in measuring university students' social skills, you could make video recordings of them as they interacted with another student whom they are meeting for the first time. Then you could have two or more observers watch the videos and rate each student's level of social skills. To the extent that each participant does in fact have some level of social skills that can be detected by an attentive observer, different observers' ratings should be highly correlated with each other. Inter-rater reliability would also have been measured in Bandura's Bobo doll study. In this case, the observers' ratings of how many acts of aggression a particular child committed while playing with the Bobo doll should have been highly positively correlated. Interrater reliability is often assessed using **Cronbach's** α when the judgments are quantitative or an analogous statistic called Cohen's κ (the Greek letter kappa) when they are categorical.

Validity is the extent to which the scores from a measure represent the variable they are intended to. But how do researchers make this judgment? We have already considered one factor that they take into account—reliability. When a measure has good test-retest reliability and internal consistency, researchers should be more confident that the scores represent what they are supposed to. There has to be more to it, however, because a measure can be extremely reliable but have no validity whatsoever. As an absurd example, imagine someone who believes that people's index finger length reflects their self-esteem and therefore tries to measure self-esteem by holding a ruler up to people's index fingers. Although this measure would have extremely good test-retest reliability, it would have absolutely no validity. The fact that one person's index finger is a centimetre longer than another's would indicate nothing about which one had higher self-esteem. Discussions of validity usually divide it into several distinct "types." But a good way to interpret these types is that they are other kinds of evidence—in addition to reliability—that should be taken into account when judging the validity of a measure. Here we consider three basic kinds: face validity, content validity, and criterion validity.

1. *Face Validity*

is the extent to which a measurement method appears "on its face" to measure the construct of interest. Most people would expect a self-esteem questionnaire to include items about whether they see themselves as a person of worth and whether they think they have good qualities. So a questionnaire that included these kinds of items would have good face validity. The finger-length method of measuring self-esteem, on the other hand, seems to have nothing to do with self-esteem and therefore has poor face validity. Although face validity can be assessed quantitatively—for example, by having a large sample of people rate a measure in terms of whether it appears to measure what it is intended to—it is usually assessed informally. Face validity is at best a very weak kind of evidence that a measurement method is measuring what it is supposed to. One reason is that it is based on people's intuitions about human behaviour, which are frequently wrong. It is also the case that many established measures in psychology work quite well despite lacking face validity. The Minnesota Multiphasic Personality Inventory-2 (MMPI-2) measures many personality characteristics and disorders by having people decide whether each of over

567 different statements applies to them—where many of the statements do not have any obvious relationship to the construct that they measure. For example, the items "I enjoy detective or mystery stories" and "The sight of blood doesn't frighten me or make me sick" both measure the suppression of aggression. In this case, it is not the participants' literal answers to these questions that are of interest, but rather whether the pattern of the participants' responses to a series of questions matches those of individuals who tend to suppress their aggression.

2. Content Validity

It is the extent to which a measure "covers" the construct of interest. For example, if a researcher conceptually defines test anxiety as involving both sympathetic nervous system activation (leading to nervous feelings) and negative thoughts, then his measure of test anxiety should include items about both nervous feelings and negative thoughts. Or consider that attitudes are usually defined as involving thoughts, feelings, and actions toward something. By this conceptual definition, a person has a positive attitude toward exercise to the extent that he or she thinks positive thoughts about exercising, feels good about exercising, and actually exercises. So to have good content validity, a measure of people's attitudes toward exercise would have to reflect all three of these aspects. Like face validity, content validity is not usually assessed quantitatively. Instead, it is assessed by carefully checking the measurement method against the conceptual definition of the construct.

3. Criterion Validity

It is the extent to which people's scores on a measure are correlated with other variables (known as criteria) that one would expect them to be correlated with. For example, people's scores on a new measure of test anxiety should be negatively correlated with their performance on an important school exam. If it were found that people's scores were in fact negatively correlated with their exam performance, then this would be a piece of evidence that these scores really represent people's test anxiety. But if it were found that people scored equally well on the exam regardless of their test anxiety scores, then this would cast doubt on the validity of the measure. A criterion can be any variable that one has reason to think should be correlated with the construct being measured, and there will usually be many of them. For example, one would expect test anxiety scores to be negatively correlated with exam performance and course grades and positively correlated with general anxiety and with blood pressure during an exam or imagine that a researcher develops a new measure of physical risk taking. People's scores on this measure should be correlated with their participation in "extreme" activities such as snowboarding and rock climbing, the number of speeding tickets they have received, and even the number of broken bones they have had over the years. When the criterion is measured at the same time as the construct, criterion validity is referred to as concurrent validity; however, when the criterion is measured at some point in the future (after the construct has been measured), it is referred to as predictive validity (because scores on the measure have "predicted" a future outcome). Criteria can also include other measures of the same construct. For example, one would expect new measures of test anxiety or physical risk taking to be positively correlated with existing measures of the same constructs. This is known as convergent validity. Assessing convergent validity requires collecting data using the measure. Researchers John Cacioppo and Richard Petty did this when they created their self-report Need for Cognition Scale to measure how much people value and engage in thinking (Cacioppo and Petty, 1982). In a series of studies, they showed that people's scores were positively correlated with their scores on a standardized academic achievement test, and that their scores were negatively correlated with their scores on a measure of dogmatism (which represents a tendency toward obedience). In the years since it was created, the Need for Cognition Scale has been used in literally hundreds of studies and has been shown to be correlated with a wide variety of other variables, including the effectiveness of an advertisement, interest in politics, and juror decisions (Petty, Briñol, Loersch, and McCaslin, 2009).

4. *Discriminant Validity*

On the other hand, is the extent to which scores on a measure are not correlated with measures of variables that are conceptually distinct. For example, self-esteem is a general attitude toward the self that is fairly stable over time. It is not the same as mood, which is how good or bad one happens to be feeling right now. So people's scores on a new measure of self-esteem should not be very highly correlated with their moods. If the new measure of self-esteem were highly correlated with a measure of mood, it could be argued that the new measure is not really measuring self-esteem; it is measuring mood instead. When they created the Need for Cognition Scale, Cacioppo and Petty also provided evidence of discriminant validity by showing that people's scores were not correlated with certain other variables. For example, they found only a weak correlation between people's need for cognition and a measure of their cognitive style—the extent to which they tend to think analytically by breaking ideas into smaller parts or holistically in terms of "the big picture." They also found no correlation between people's need for cognition and measures of their test anxiety and their tendency to respond in socially desirable ways. All these low correlations provide evidence that the measure is reflecting a conceptually distinct construct.

REFERENCES

C.R. Kothari, Research Methodology, Methods and Techniques, Wiley Eastern Limited, New Delhi, 1988.

Claire Sellitiz and others, Research Methods in Social Sciences, 1962, p.50

Cacioppo, J. T., and Petty, R. E. (1982). The need for cognition. Journal of Personality and Social Psychology, 42, 116–131.

Dollard, J., Criteria for the Life-history, Yale University Press, New York,1935, pp.8-31.

L.V. Redman and A.V.H. Mory, The Romance of Research, 1923.

Marie Jahoda, Morton Deutsch and Staurt W. Cook, Research Methods in Social Relations, p.4.

Panneerselvam, R., Research Methodology, Prentice hall of India, New Delhi, 2004.

Pauline V. Young, Scientific Social Surveys and Research, p.30

Petty, R. E, Briñol, P., Loersch, C., and McCaslin, M. J. (2009). The need for cognition. In M. R. Leary and R. H. Hoyle (Eds.), Handbook of individual differences in social behaviour (pp. 318–329). New York, NY: Guilford Press.

T.S. Wilkinson and P.L. Bhandarkar, Methodology and Techniques of Social Research, Himalaya Publishing House, Bombay, 1979.

The Encyclopedia of Social Sciences, Vol. IX, Macmillan, 1930.

— *Chapter 2* —

Analytical Methods for Research Data

Statistics is a branch of science which deals with the collection, classification, description and interpretation of data obtained by conducting surveys and experiments. The essential purpose of it is to describe and draw valid inferences about numerical properties of populations" (Ferguson 1966). The experience in statistics application indicates that a single statistical method can be used in different research areas for dealing with different types of problems. Various possibilities which application of a statistical method provides in certain research areas should be considered as an adequate way of dealing with a problem of the research. However, it should be noted that statistics is not a method by which it is possible to solve all the problems in a research. Nachmias and Nachmias (2009) stated that there are basically two types of statistics which include descriptive and inferential statistics. Descriptive statistics enable the researcher to summarize and organize data in an effective and meaningful way. It involves the use of tables, charts, graphs, mean, modes, median, standard scores and correlation to treat collected data. Inferential statistics is concerned with making inferences from a unit of a population. Inferential statistics is concerned with making inferences from a unit of a population. Inferential statistics allow the researcher to make decisions or inferences by interpreting data patterns. Researchers use inferential statistics to determine whether an expected pattern designated by the theory and hypotheses is actually found in the observations. To decide whether this hypothesis is true, researchers might survey the respondents and then use descriptive statistics to make comparison between these groups and would employ inferential statistics to determine whether the differences between the groups are significant or not.

Social research is a type of research conducted by researchers following a systematic plan. Social research methodologies can be classified along a quantitative/qualitative dimension. Quantitative designs approach social phenomena through quantifiable evidence, and often rely on statistical analysis of many cases (or across intentionally designed treatments in an experiment) to create valid and reliable general claims. Qualitative designs emphasize understanding of social phenomena through direct observation, communication with participants, or analysis of texts, and may stress contextual subjective accuracy over generality. When applying a statistical method, it is common to differentiate between quantitative and qualitative features and variables. Nominal and ordinal variables and data are usually considered as qualitative (attributive), while interval variables and ratio variables are considered as quantitative (Ferguson 1966, Krneta 1987). Also, it is common to apply nonparametric statistical methods on nominal and ordinal data, while parametric methods are used for the interval and ratio data (Ferguson 1966).

Statistics and statistical methods have highly significant application in sociology. Functions of statistics are numerous: the methods of descriptive statistics have an important application for describing natural phenomena; inferential statistics is used for inductive reasoning about unknown properties of a larger group using the known indicators of the causes; hypothesis testing most frequently refers to the results of one, two or more causes, on the basis of which it is possible to draw conclusions on the problem of the research, by accepting or refuting an initial hypothesis; regression and correlation analysis, in the most simple case, examines the influence and dependence between two or more variables.

2.1 Parametric Methods/Tests

The two classes of statistical tests are called parametric and nonparametric. The word parametric comes from "metric," meaning to measure, and "para," meaning beside or closely related; the combined term refers to the assumptions about the population from which the measurements were obtained. Nonparametric data do not meet such rigid assumptions. Nonparametric tests sometimes are referred to as "distribution-free." That is, the data can be drawn from a sample that may not follow the normal distribution.

Before a parametric test can be undertaken, it must be ascertained that: 1) the samples are random (*i.e.*, each member of the population has an equal chance of being selected for measurement); 2) the scores are independent of each other; 3) the experiments are repeatable with constancy of measurements from experiment to experiment; 4) the data are normally distributed; and 5) the samples have similar variances (1, 2). Parametric statistics use mean values, standard deviation and variance to estimate differences between measurements that characterize particular populations.

The two major types of parametric tests are Student's t-tests and analyses of variance (ANOVA). Nonparametric tests use rank or frequency information to draw conclusions about differences between populations. Parametric tests usually are assumed to be more powerful than nonparametric tests. However, parametric tests cannot always be used to analyze the significance of differences because the assumptions on which they are based are not always met.

2.1.1 Basic Concepts of Testing of Hypothesis

a) Hypothesis

A *hypothesis* is a statement about the parameters of the underlying distribution of the observations (specified in the model). If parameters are unknown for any test, we do not know whether a hypothesis is true or false. But data are evidence, and we might have enough evidence against a hypothesis to allow us to reject it.

Any testing involves two hypotheses: the *null hypothesis*, H_0, and *the alternative hypothesis*, H_1. The null hypothesis is protected: it will only be rejected if there is strong evidence against it in favour of H_1.The alternative is only a benchmark to test against. The null hypothesis is what you want to prove false. The result of a statistical investigation is given in terms of the null hypothesis: if there is insufficient evidence against H_0 we say "we do not reject H_0" or "we accept H_0"; otherwise we say "we can reject H_0".

Examples

☆ (Biological researcher) H_0: $\theta = 0.3$, H_1: $\theta > 0.3$

☆ (Weight (in Kg.) observer) H_0: $\mu = 5$, H_1: $\mu < 5$

b) Errors

A type I error occurs if we reject H_0 when it is true. This is a serious error. The probability of making a type I error should be low. A *type II error* occurs if we fail to reject H_0 when it is in fact false. This is a less serious error.

c) Test Statistics and Critical Regions

A test consists of a test statistic, such as the sample mean, sample variance or something more complicated and a critical region, specifying which values of the test statistic will result in rejection of H_0.

Examples

☆ (Biological researcher) Reject H_0 if more than half the patients recover

- ☆ (Weight observer) Reject H_0 if the sample mean weight loss is no more than 3.1kg.
- ☆ Biased coin.

d) Significance and P-values

The distribution of the test statistic (the *reference distribution*) is determined by the model; the parameter value is specified by the hypothesis. So we can find the cut off point for the critical region which gives a type I error probability of exactly 5 per cent, say.

If the observed value of the test statistic falls in this critical region, we say it is *significant* and then we reject H_0 *at the 5 per cent level.*

We can do the same for 1 per cent (*highly significant*) and 0.1 per cent (*very highly significant*) as well. The p-value measures the significance level at which the observed value of the test statistic first becomes significant. In other words, it answers the question "How unlikely is this result if H_0 is true?"

e) One-sided and Two-sided Tests

Whether a test is one-or two-sided depends on the alternative H_1. An alternative involving ">" or "<" is one-sided; one involving both is two-sided. (The words *one-tailed* and *two-tailed* are also used).The critical region for a two-sided test at 5 per cent significance level involves 2½ per cent in each tail of the distribution. For this reason, when you are dealing with a two-sided test, the p-value which you calculate from one-sided tables must be doubled.

2.1.2 Different Types of Parametric Methods/Tests

a) Testing the Mean when the Variance is known

Since the sample mean has $N(\mu, \delta^2/n)$ distribution, we can define

$$Z = \frac{\overline{x} - \mu}{\frac{\delta}{\sqrt{n}}}$$

and state that Z ~ N(0,1).Now δ is known, and μ is specified by H_0, so we can evaluate Z from the data *under the assumption that H_0 is true.*

b) Testing the Mean when the Variance is unknown

The value of Z used above still has N (0, 1) distribution. But it is no longer a statistic, as s is unknown. We must use $S^2$to estimate s^2 to give:

$$T = \frac{\overline{x} - \mu}{\frac{S}{\sqrt{n}}}$$

T is not Normal; it has a *Student's t* distribution on n - 1 degrees of freedom. The testing procedure is as above, but using T instead of Z and finding the cut off points of the critical regions in t tables instead of Normal tables.

Large Sample Approximation

When n is large the t distribution approaches the Normal. t tables usually stop at 120 degrees of freedom; for $n > 120$ it is acceptable to use Normal instead of t. Some books advocate a lower cut off point, such as 30 d.f.

Degrees of Freedom

The term "degrees of freedom" makes sense in the context of testing goodness of fit. Loosely, start with n d.f., then lose 1 for each unknown parameter estimated in the model for the mean.

c) Testing the Variance

If $Z_1, .., Z_n$ are independent standard Normal variables, then

$$\sum_{i=1}^{n} Z_i^2 \sim \chi_n^2$$

χ_n^2 is the *chi-squared* distribution on n degrees of freedom.

Therefore, for $X_1. X_n \sim, N(\mu, \delta^2)$

$$\Sigma \; \frac{(X_i - \mu)^2}{\delta^2} \sim \chi_n^2$$

If μ is known, we can use a test based on χ_n^2. Usually μ is unknown and must be estimated.

$$\frac{(X_i - \overline{X})^2}{\delta^2} \sim \lambda_{n-1}^2$$

Note that the LHS is $\frac{(n-1)S^2}{\delta^2}$. In addition; sample variance and sample mean are independent random variables. To test H_0: $\delta^2 = 16$ (say) against H_1: $\delta^2 > 16$, calculate $(n\text{-}1)\, S^2/16$ and compare with λ_{n-1}^2 tables: if the statistic is too large, reject the null hypothesis.

Robustness

All tests are based on a model. Some work well even when the assumptions of the model are not satisfied (eg, underlying distribution is not really Normal): they are *robust tests*. The t-test is robust. The test based on λ^2 is not.

d) Comparative Tests Based on Two Normal Samples

We assume that there are two independent simple random samples X_1.., $X_n \sim N(\mu_x, \delta_x^2)$ and Y_1,.., $Y_m \sim N(\mu_y, \delta_y^2)$. We may want to test whether $\mu_x = \mu_y$ or whether $\delta_x^2 = \delta_y^2$

i) *Two-sample t Test*

Here we test H_0: $\mu_x = \mu_y$ against H_1: $\mu_x \neq \mu_y$ (or a one-sided alternative).

The test presented only works if $\delta_x^2 = \delta_y^2$. (Later we see how to test this.)

Let δ^2 denote the common variance. Then, if H_0 is true,

$$\overline{X} - \overline{Y} \sim N\left[0, \delta^2\left(\frac{1}{n} + \frac{1}{m}\right)\right]$$

which allows us to deduce that Z ~ N(0,1). In general δ is unknown and must be estimated. We use a combination of S^2_X and S^2_Y, called the *pooled estimate of the variance,*

$$S^2p = \frac{(n-1)S^2{}_X + (m-1)S^2{}_Y}{n+m-2}$$

Now we can say that t ~ t_{m+n-2} and use a t test.

ii) *Paired Sample Test*

Here the two samples have the same size and there is a natural correspondence between individual observations from the two samples. For example, the before and after test is paired sample test. In these cases we calculate the differences, After - Before, or the ratios, After/Before, and treat that as a single sample. A t test will usually be appropriate.

e) Testing Equality of Variance

We test H_0: $\delta_x^2 = \delta_y^2$ against H_1: $\delta_x^2 \neq \delta_y^2$ (two-sided alternative).The test used is the F test. If H_0 is true, then the ratio S_X^2/S_Y^2 has an F distribution with (n-1, m-1) degrees of freedom.

It is equally true to say that $S_Y^2/S_X^2 \sim F_{m-1, n-1}$.Because of the way the F tables are organized, we always put the largest sample variance on top.

What if Variances aren't Equal?

Various possibilities: Welch test (distribution is approximately t), Behrens' test (exact, but tables are hard to read), or a non-parametric test such as the Wilcoxon rank sum test (see later).

f) Tests Based on Binomial Samples

i) Single Binomial Sample

Example: A course review claims that on average at least 60 per cent of the students end up with a first or upper second. Of the 60 graduates in the last 3 years, 32 have been awarded firsts or II-1s. Does this contradict the claims of the review?

Basic idea as before: specify a model, write down the hypotheses, find a test statistic, calculate its distribution under H_0, and look at the observed value to see how unusual it is. In the example the model is that each student, independently, gets a first or II-1 with probability θ.Hypotheses are H_0: θ= 0.6, H_1: θ < 0.6. Test statistic, *X*, is the number of students who got firsts and II-1s.If H_0 is true, *X* ~ Bin (60, 0.6).This is roughly N (36, 14.4), so define

$$Z = \frac{X-36}{\sqrt{14.4}} = -1.054$$

Not significant.

ii) Comparison of Two Binomial Samples

Here we are testing equality of proportions. The model is that each of the n_X units in the first sample has probability θ_x f possessing a particular property; each of the n_Y in the second sample has probability θ_y.Usually H_0 is that $\theta_x = \theta_y$ with a 1-sided or 2-sided alternative.

If H_0 is true, then

$$\frac{X}{n_X} - \frac{Y}{n_Y} \text{ has mean } 0,$$

$$\text{variance } \theta(1-\theta)\left(\frac{1}{n_X} + \frac{1}{n_Y}\right)$$

where θ is the common value of θ_X and θ_Y. But θ is unknown, so must be estimated using both samples: we use

$$\frac{X+Y}{n_X + n_Y}$$

With this estimate of θ we employ a Normal approximation. Tests using Normal approximations can be used in more cases than just Binomial, though they are not the "best" tests.

A statistical test can never establish the truth of a hypothesis with 100-percent certainty. Typically, the hypothesis is specified in the form of a "null hypothesis," *i.e.*, the score characterizing one group of measurements does not differ (within an acceptable margin of measurement error) from the score characterizing another group. Note the hypothesis does not state the two scores are the same; rather, it states no significant difference can be detected. Performing the statistical procedure yields a test result that helps one reach a decision that the scores are not different (the hypothesis is confirmed) or the difference in the scores is too great to be explained by chance (the hypothesis is rejected).Rejecting the hypothesis when it actually is true is called a Type-I error. Failure to reject the hypothesis when it is false is termed a Type-II error. For convenience and simplicity, a 5-percent risk of making a Type-I error has become conventional; one should be correct 95 out of 100 times when using the listed value in the probability tables to accept or reject the hypothesis. In statistics, robustness is the degree to which a test can stray from the assumptions before changing the confidence you have in the result of the statistical test you have used. Choosing a nonparametric test trades the power, or large sample size, of the parametric test for robustness. Further, a method requiring few and weak assumptions about the population(s) being sampled is less dependable than the corresponding parametric method and increases the chances of committing a Type-Il error.

2.1.3 Independence or Dependence of Samples

Independence or dependence of samples concerns whether the different sets of numbers being compared are independent or dependent of each other. Sets are independent when values in one set tell nothing about values in another set. When two or more groups consist of different, unrelated individuals, the observations made about the samples are independent. When the sets of numbers consist of repeated measures on the same individuals, they are said to be dependent. Similarly, if male and female characteristics are compared using brother-sister pairs, the samples are dependent. Matching two or more groups of individuals on factors such as income, education, age, height and weight also yields dependent samples.

2.1.4 Random Selection from Normally Distributed Populations

This type of selection sometimes is difficult to confirm. However, so long as the data sets used in the analysis are relatively normally distributed, the robustness of most parametric tests still provides an appropriate level of rejection of the null hypothesis.

2.1.5 Homogeneity of Variance

Homogeneity of variance of the data from each group being compared must be equal (homogeneous) and can be tested statistically. If it is found to differ significantly, then nonparametric tests must be used.

Parametric tests require data from which means and variances can be calculated, *i.e.*, interval and ratio data. Some statisticians also support the use of parametric tests with ordinal-scaled values because the distribution of ordinal data often is approximately normal. As long as the actual data meet the parametric assumptions, regardless of the origin of the numbers, then parametric tests can be conducted. As is the case with all statistical tests of differences, the researcher must interpret parametric statistical conclusions based on ordinal data in light of their clinical or practical implications.

2.2 Non Parametric Methods

Besides the Parametric methods, another method commonly used in statistics to model and analyze ordinal or nominal data with small sample sizes. Unlike parametric models, nonparametric models do not require the modeler to make any assumptions about the distribution of the population, and so are sometimes referred to as a distribution-free method.

2.2.1 Uses of Nonparametric Tests

Nonparametric tests are used in the behavioural sciences when there is no basis for assuming certain types of distributions. Siegel has advocated that nonparametric tests be used for nominal and ordinal levels of measurements while parametric tests are used for analyzing interval and ratio data.

On the other hand, Williamson has argued that statistical tests are selected to meet certain goals or to answer specific questions rather than to match certain levels of measurement with parametric or nonparametric procedures. This view currently is prevalent among many statisticians.

In practice, levels of measurement sometimes are "downgraded" from ratio and interval scales to ordinal or nominal scales for the convenience of a measuring instrument or interpretation. For example, muscular strength (measured with a force gauge and considering the length of the lever arm through which the force is acting) is a variable that yields ratio data because a true zero point exists in the level of measurement. Muscular strength is absent with paralysis (true zero point). The manual muscle test converts the ratio characteristic of force into an ordinal scale by assigning grades of relative position (normal, good, fair, poor, trace, zero; or 5, 4, 3, 2, 1, 0).

Nonparametric tests should not be substituted for parametric tests when parametric tests are more appropriate. Nonparametric tests should be used when the assumptions of parametric tests cannot be met, when very small numbers of data are used, and when no basis exists for assuming certain types or shapes of distributions.

Nonparametric tests are used if data can only be classified, counted or ordered-for example, rating staff on performance or comparing results from manual muscle tests. These tests should not be used in determining precision or accuracy of instruments because the tests are lacking in both areas.

2.2.2 Advantages

Nonparametric tests usually can be performed quickly and easily without automated instruments (calculators and computers). They are designed for small numbers of data, including counts, classifications and ratings. They are easier to understand and explain.

Calculations of nonparametric tests generally are easy to perform and apply, and they have certain intuitive appeal as shortcut techniques. Nonparametric tests are relatively robust and can be used effectively for determining relationships and significance of differences using behavioral research methods.

2.2.3 Disadvantages

Parametric tests are more powerful than nonparametric tests and deal with continuous variables whereas nonparametric tests often deal with discrete variables. Using results from analyses of nonparametric tests for making inferences should be done with caution because small numbers of data are used, and no assumptions about parent populations are made. The ease of calculation and reduced concern for assumptions have been referred to as "quick and dirty statistical procedures".

2.2.4 Nonparametric Methods for Descriptive Statistics

Descriptive statistics involve tabulating, depicting and describing collections of data. These data may be either quantitative, such as measures of leg length (variables that are characterized by an underlying continuum) or representative of qualitative variables, such as gender, vocational status or personality type.

Collections of data generally must be summarized in some fashion to be more easily understood. Descriptive statistics serve as the means to describe, summarize and reduce to manageable form the properties of an otherwise unwieldy mass of data. Descriptive statistics used to characterize data analysed by parametric tests include the mean, standard deviation and variance. Those descriptive statistics used to characterize data analyzed by nonparametric tests include the mode, median and percentile rank:

- ✰ Mode is the most frequently occurring score within a distribution. It is possible to have a sample of scores that has no mode, and it is possible to have two or more modes; if there are two modes, the distribution is called bimodal.
- ✰ Median is the middle score of a ranked distribution of measurements.
- ✰ Percentile or percentile rank of an observation is the percentage of a distribution that falls below that observation. Percentile rank is calculated with below Equation:

$$PR_i = 100(1 - \frac{R_i - 0.5}{n})$$

where R_i is the rank of the observation X_i (ranked from highest to lowest), and n is the number of observations in the distribution. The median is the 50th percentile.

In statistics, the mean or median commonly is used when dealing with measurement data. The mode most often is useful when dealing with data more appropriately handled with classification procedures (*e.g.*, mild, moderate, severe).

2.2.5 Measures of Association (Correlation)

Correlation coefficients are used to reveal the nature and extent of association between two variables. Each method used to determine a correlation coefficient has conditions that must be met for its use to be appropriate. The first step in analysing a relationship always is selection of the proper measure of association based on the conditions of the study and the hypothesis to be tested.

Measures of association are useful for a variety of studies. Correlation coefficients are used in exploratory studies to determine relationships among variables in new study areas. The results of such studies allow investigators to formulate further research questions or hypotheses to delve more deeply into the study area. In some studies, the hypotheses focus on associations between selected variables, and the correlation coefficients serve to test these hypotheses.

Similarly, hypotheses based on expected associations among variables make important contributions to theory building.

Finally, correlation coefficients are used to manage threats to validity in experimental and quasi-experimental studies. They can be used to test the credibility of findings when groups have been compared by checking on the association of independent and extraneous variables with the dependent variable.

2.2.6 Spearman's Rank Order Correlation Coefficient

Spearman's rank order correlation coefficient *rho* is a nonparametric method of computing correlation from ranks. The method is similar to that used to compute Pearson's correlation coefficient (a parametric test), with the computed value *rho* providing an index of relation between two groups of ranks.

If the original scores are ranks, the computed index will be similar in value to that computed by the Pearson (product moment) method. The is the product moment method assigns weight to the magnitude of each score whereas the rank method focuses on the ordinal position of each score. The coefficient of rank correlation (*rho*) ranges from +1, when paired ranks are changing in the same order, to -1, when ranks are changing in reverse order. A score of zero indicates the paired ranks are occurring at random. The equation for rank correlation is:

$$\rho = 1 - \frac{6\sum d_i^2}{n(n^2 - 1)}$$

where d is the difference between each subject's rank for the two variables being studied, Σd^2 is the sum of squared differences between ranks, 6 is a constant and n is the number of paired scores.

Suppose 10 students are drawn at random from a large class; each student has been rated on a 10-point scale for a recent clinical experience, and each student has a grade-point average (GPA) on file. The coefficient of ranks can be computed to determine the extent of agreement between the two sets of scores (clinical experience ratings and GPA). In the rank correlation method, the raw scores are replaced by assigning an appropriate rank to each score of each set. Ranks for each set correspond to the total number of scores in that set.

Step 1

Make a table of the subjects' scores and ranks for the two variables of interest and subtract the ranks to determine the difference (diff) for each pair of ranks. Square each of these differences and sum the squared values.

This example illustrates what happens when scores are similar (tied ranks). When tied ranks occur, each score is assigned the average rank the tied scores occupy (a higher rank is better). The GPA of 3.2, for example, had two scores occupying ranks 5 and 6. The average rank for the score 3.2 is obtained by adding ranks (5 +6) and dividing by the number of ranks occupied (*e.g.*, 5 +6 + 2= 5.5 ranking).

Step 2

Substitute the calculated value of Σd^2 and solve for ρ:

$$\rho = 1 - \frac{6\sum d_i^2}{n(n^2 - 1)} = 1 - \frac{(6)\cdot(22)}{1000 - 10} = 0.867$$

Consulting a textbook of statistics that provides a table of values for dilu123, one finds a minimum dilu123 value of 0.746 is needed to be considered significant at the 0.05 level of significance. Thus, the correlation coefficient dilu123 of 0.867 confirms a statistically significant correlation between the two sets of rankings.

2.2.7 Kendall's Rank Correlation

Kendall's rank correlation tau ('r) is another nonparametric measure of association. When relatively large numbers of ties exist in a set of ranking, Kendall's tau is preferred over Spearman's *rho*. The formula and procedures for calculating it have been adapted from Siegel (33).

$$\tau = \frac{S}{(N/2)(N-1)} \text{ =Actual score/Maximum possible score}$$

where N = the number of objects or individuals ranked on both X and Y characteristics.

The value of S can be determined by arranging the first set of measurements into their natural order (*e.g.*, 1, 2, 3, 4, 5) and aligning the second set of measurements under them (*e.g.*, 2, 1,4, 5, 3). Starting with the first number on the left in the bottom row, the number of ranks on the right which are larger are counted. The derivations of the actual score and the maximum possible score are illustrated in the example that follows: Two orthotists rank the fit of an "off-the-shelf" ankle-foot orthosis (AFO) on five different patients.

Step 1

Rearrange the data so the first orthotist's rankings fall in a natural (increasing) order and the second orthotist's rankings are tabulated in the same order.

Step 2

Compare the first ranking of orthotist 2 with every ranking to its right, assigning a +1 to every pair in which the order is natural and a -1 to every pair in which the order is unnatural:

- ☆ Comparing 2 with 1 produces a -1
- ☆ Comparing 2 with 4 produces a +1
- ☆ Comparing 2 with 5 produces a +1
- ☆ Comparing 2 with 3 produces a +1

Repeat for each subsequent ranking of orthotist 2:

- ☆ comparing 1 with 4 produces a +1
- ☆ comparing 1 with 5 produces a +1
- ☆ comparing 1 with 3 produces a +1
- ☆ comparing 4 with 5 produces a +1
- ☆ comparing 4 with 3 produces a -1
- ☆ comparing 5 with 3 produces a -1
- ☆ The sum of the comparisons: 4

Step 3

Add these measures of "disarray" (sum = 4) and enter this sum in the above formula as a substitute for S.

Step 4

The value of N = 5. Thus, it becomes:

$$\tau = \frac{S}{(N/2)(N-1)} = \frac{4}{(2.4).(4)} = 0.416$$

Step 5. The statistical significance can be determined by two procedures, depending on sample size.

If N is equal to or less than 10, use a probability table such as that found in the appendix of a textbook on statistics to find the statistical significance of 'r. In this example, the table of probability indicates a probability score (p-value) of 0.242 for a value of 0.400. Thus, this test supports the conclusion that the ratings of the two orthotists are not significantly correlated.

For situations in which N is greater than 10, a z score can be computed for the 'r obtained and the statistical significance of the correlation read from a corresponding table of z scores:

$$z = \frac{\tau}{\sqrt{\frac{2(2N+5)}{9N(N-1)}}}$$

2.2.8 Chi-square Test of Independence

The Chi-square test of independence is a nonparametric test designed to determine whether two variables are independent or related. This test is designed to be used with data that are expressed as frequencies; it should not be used to analyze data expressed as proportions (percentages) unless they are first converted to frequencies.

The application of Chi-square to contingency tables can best be illustrated by working through an example. Suppose a sample of new graduates of an orthotic educational program and orthotists with more than five years of clinical experience were asked whether research should be a part of every orthotist's practice. The replies were recorded as "Agree" or "Disagree."

Step 1

Organize the data into the form of a 2 x 2 contingency table. Note the table includes row totals, column totals and the grand total of subjects included in the sample.

The actual numbers of "Agree" responses were 82 from recent graduates and 30 from experienced orthotists. The numbers disagreeing with the statement were 12 and 66, respectively.

The rationale that underlies Chi-square is based on the differences between the observed and the expected frequencies. The observed frequencies are the data produced by the survey. The expected frequencies are computed on the assumption that no difference existed between the groups except that resulting from chance occurrences.

Step 2

The expected frequencies are computed as follows:

$$\frac{(a+b)}{n}\times(a+c)=\frac{(82+12)}{190}\times(82+30)=55.41 = \text{Expected } a'$$

$$\frac{(a+b)}{n}\times(b+d)=\frac{(82+12)}{190}\times(12+66)=38.59 = \text{Expected } b'$$

$$\frac{(c+d)}{n}\times(a+c)=\frac{(30+66)}{190}\times(82+30)=56.54 = \text{Expected } c'$$

$$\frac{(c+d)}{n}\times(b+d)=\frac{(30+66)}{190}\times(12+66)=39.41 = \text{Expected } d'$$

Cross-tabulation and the computation of Chi-square can be made when the variables are nominal as well as ordinal, interval or ratio, and the Chi-square statistic is useful for discrete or continuous variables. However, it is assumed that data occur in every category; thus, no cell may have an observed frequency of zero. The formula for the degrees of freedom for calculating Chi-square and the contingency coefficient is:

$df=(k-1)(r-1)$ (5)

where k = number of columns in the contingency table and r = number of rows in the contingency table.

Step 3

The Chi-square (X^2) is calculated as below:

$$\chi^2=\sum_{i=1}^{k}\frac{(O_i-E_i)^2}{E_i}$$

where O is the observed number of cases found in the ith row of the ith column, and E is the expected frequency obtained by multiplying the two marginal totals for each cell and dividing the product by N

Step 4

The Chi-square is computed by finding the difference between the observed and expected frequencies in each cell, squaring that difference and dividing by the expected frequency of that cell. The result for each cell is then added, and the total is the value of the Chi-square. (Chi-square = X^2 = 61.84.)

Step 5

Consulting a table of Chi-square values in a textbook of statistics, using 1 degree of freedom and the 0.05-level of significance, we find that a minimum value of 3.84 is needed for the observed frequency to be considered significantly different from the expected frequency. In this example, the value of X greatly exceeds that minimum value; thus, the observed values are significantly different from the expected values.

The use of the Chi-square statistic has important limitations. Although no association is indicated by a zero, a perfect association is not indicated by a 1.00. Moreover, the size of Chi-square is influenced by both the size of the contingency table and the size of the sample.

The addition of rows and columns as a table grows is accompanied by larger and larger values of Chi-square- even when the association remains essentially constant. If the sample size is tripled, the value of Chi-square is tripled, and everything else remains the same. Degrees of freedom depend on the number of rows and columns, not the sample size; thus, inflated values of Chi-square occur for large samples, leading the investigator to conclude the differences between observed and expected frequencies are more significant than warranted. The Chi-square is designed for use with relatively small samples and a limited number of rows and columns.

2.2.9 Phi

The correlation coefficient *phi* corrects for the size of the sample when the table size is 2 x 2. The equation is:

$$\phi = \sqrt{\frac{\lambda^2}{N}}$$

Phi is 0 when no relationship exists and 1 when variables are related perfectly. When tables are greater than 2 x 2, *Phi* has no upper limit and is not a suitable statistic to use. The statistical significance of *Phi* may be tested by calculating a corresponding Chi-square value and assigning 1 degree of freedom to it:

$$\lambda^2 = N(\phi)$$

2.2.10 Cramer's V

Cramer's V is an adjusted CF, modified to be suitable for tables larger than 2 x 2. The value of V is zero when no relationship exists and 1 when a perfect relationship exists. The equation for Cramer's V is:

$$V = \sqrt{\frac{\lambda^2}{N(K-1)}}$$

Thus, when 2 x 2 tables are involved, Phi may provide a more useful measure of the relationship between the two variables than that provided by Chi-square. For tables larger than 2 x 2, Cramer's V is the statistic of choice (k = number of columns in the table).

Tests for Significance of Differences

2.2.11 Two-Group Design: Chi-square (2 x 2)

The Chi-square comparison of differences between two groups is one of the better known and commonly used statistical procedures. The same procedure for Chi-square (x^2) as described above can be used to test for the significance of differences in two groups of data that are expressed as frequencies.

Suppose researchers wanted to determine if the proportion of trauma patients being referred for orthotic services in a particular hospital was significantly different than the number being referred for orthotic services in another hospital with a similar mix of patients. During a specific 12-month period, the orthotic department in Hospital A filled 238 requests for orthoses from a pool of 2,222 patients, and the orthotic department in Hospital B filled 221 requests for orthoses from a pool of 1,238 patients. First, the data are organized into a 2 x 2 contingency table.

As before, the general equation for Chi-square is:

$$\chi^2 = \sum_{i=1}^{k} \frac{(O_i - E_i)^2}{E_i}$$

To compute X for a contingency table, simply square the difference between the observed and expected frequencies in each cell and divide by the expected frequency of that cell. Finally, total the cells to obtain the X^2value. X^2= 23.72, which is evidence that the experiences of the two hospitals are significantly different.

2.2.12 Two-Group Design: Chi-square Median Test

The Chi-square median test can be used to determine if the medians of two groups are different. For example, all of the male patients fitted with an AFO to correct foot-drop following the onset of hemiplegia were asked to rate the comfort of their footwear when walking with the AFO. Forty-four patients were evaluated; 32 wore normal leather shoes and 12 wore tennis shoes.

Comfort was rated on a nine-point scale (the larger the score, the greater the comfort in walking), and the evaluation was made six months after fitting the AFO and with the subject walking 50 yards. The median comfort rating of the 44 patients was 7.3. The number of subjects rating their comfort above or below is the grand median.

The Chi-square computation viewing the leather shoe and tennis shoe wearers as random samples is shown below. The ratings are discrete units; each patient's rating appears only once, and the ratings are independent.

$$\chi^2 = n_i[\sum_{r=1}^{R} + \sum_{c=1}^{C} + \frac{n_{rc}^2}{n_r \cdot n_c} - 1]$$

where n_i the total number of observations, n_{rc}^2 is the number of observations in the rc[th] cell of the contingency table, n_r is the number of observations in the rth row of the table, and n_c is the number of observations in the cth column of the table.

$$\chi^2 = 44\left[\frac{4^2}{19\times(12)} + \frac{15^2}{19\times(32)} + \frac{8^2}{25\times(12)} + \frac{17^2}{5\times(32)} - 1\right]$$

$= 44(0.070175+0.370066+0.213333+0.361250-1)$

$= 44(1.014825-1)$

$= 0.652$

A table of Chi-square values is then consulted to determine if this calculated value of Chi-square is sufficiently large to represent a statistically significant difference of the mean scores. The degree of freedom is (R - 1) (C - 1) = (1)(1) = 1. In this example, a Chi-square of 3.84 or larger would be needed (at a.05 level of significance) to justify the conclusion that the comfort levels of the two different types of footwear were significantly different.

2.2.13 Two-Group Design: Tukey's Quick Test

Tukey's quick test is used to determine if the results of two different interventions produced the same or different effects. Suppose a sample of 20 patients with limitation in elbow extension on one side that exceeded 50 degrees was treated with one of two methods for reducing contractures. Subgroup A, consisting of 10 patients, was treated with serial casting over a period of one month; Subgroup B, also consisting of 10 patients, was treated with an adjustable splint worn 18 hours a day for one month. The increased range of motion for subjects were in the two groups.

Tukey's quick test is applied by identifying the group containing the largest value and the group containing the smallest value in the two groups. In this example, Group B contains the largest value of either group (41), and Group A contains the smallest value (14). The number of values in Group B that are larger than the largest value in Group A (36) are counted and recorded (in this example, there are 2). Next, the number of values in Group A that are smaller than the smallest value in Group B (18) are counted (there are 2). The two counts are added, and, if the sum is equal to or greater than 7, we conclude the effects of the two treatments are different. If the sum is less than 7, we conclude the effects of the two treatments are not different. In the present example, the sums of the two counts equal 4; therefore, we conclude the effects of the two interventions are not different. In the event the largest and the smallest values occurred in the same group, we conclude automatically that the two treatments did not have different effects. In Tukey's test the number 7 is a constant and is the criterion value to be used with any set of data.

2.2.14 Two-Group Design: Mann-Whitney U-Test

The Mann-Whitney U-test is a rank test for two independent samples, each with a small number of subjects. This test is a good alternative to the parametric t-test. Suppose measurements of the height of the ankle joint axis (in millimetres) in a group of patients receiving services in Orthotic Clinic A are compared with measurements taken from a group of patients in Orthotic Clinic B to determine if they are comparable. Because of the small number of cases, a nonparametric test is selected. The measurements are assigned a rank in ascending order of height, with a rank of 1 being the smallest value:

$$\chi^2 = 44\left[\frac{4^2}{19\times(12)}+\frac{15^2}{19\times(32)}+\frac{8^2}{25\times(12)}+\frac{17^2}{5\times(32)}-1\right]$$

$$= 44(0.070175+0.370066+0.213333+0.361250-1)$$

$$= 44(1.014825-1)$$

$$= 0.652$$

$$H = \frac{12}{N(N+1)}$$

The ranks then are ordered according to their identity.

The value of the Mann-Whitney U-test is found by determining the number of A scores preceding each B score. The U is: 1 + 2 +3 + 4 + 5 = 18 (rank 2A precedes 3B = +1; ranks 2A and 4A precede 5B = +2; ranks 2A, 4A and 6A precede 7B = + 3 and so on). Consulting a Mann-Whitney U-test table for nB = 7 (larger sample size), locate the U value (18) on the left-hand margin and nA = 5 at the top of the table. The probability that these two samples are equivalent is 0.562, which is not statistically significant (*i.e.*, the distribution of ankle heights is not different). This procedure is appropriate only when the larger sample size is 8 or smaller. Different procedures and tables are used for samples ranging between 9 and 20 and larger than 20, respectively. Procedures and tables for the Mann-Whitney

U-test can be found in Siegel and Castellan.

2.2.15 Two-Group Design: Wilcoxon Matched Pairs/Rank Test

The Wilcoxon matched pairs/rank test is an alternate form of the Mann-Whitney test that is used when the samples are dependent. For purposes of illustration, presume the time to ambulate 25 meters is measured with a stopwatch when the patient is wearing a new type of lightweight KAFO and again when wearing a conventional metal KAFO. The ambulation times for each patient are tabulated, and the absolute difference between each pair of numbers is calculated. The nonzero differences then are ranked according to their absolute values and separated into ranks associated with positive and negative differences.

As in the case with the Mann-Whitney procedures for analyzing differences between independent samples, the resulting score, in this case called a T value, is used to look up the statistical significance of the differences in a table. In this example, a T value of 8 or more indicates that the two situations are significantly different, and the subjects walked more quickly when wearing the lightweight KAFO.

Multi-Groups Design with Independent Samples: The Kruskal-Wallis One-Way Analysis of Variance by Ranks. This method functions like the conventional one-way analysis of variance. The null hypothesis is tested to determine if the differences among samples show true population differences or whether they represent chance variations to be expected among several samples from the same population. The test is based on the assumptions that ranks within each sample constitute a random sample for the set of numbers 1 through N (15) and that the variable being tested has a continuous distribution (4). Scores in all samples are combined and arranged in order of magnitude so that ranks can be given to each score. The lowest score is assigned the rank of 1. The scores then are replaced in their respective samples with appropriate ranks. The ranks for each sample are summed. The assumption is that mean rank sums (R) are equal for all samples and equal to the mean of the N ranks, (N + 1)12, if the samples (K) are from the same population (16). Both equal- and un-equalized samples can be used in this test because the sums of sample ranks (>R) are pooled in the equation. The statistic H used in this test can be defined by the equation:

$$H = 12/[N (N+1)]$$

where N is the number of scores in all samples combined. The random sample distribution of H is approximated by a Chi-square distribution of K-1 degrees of freedom, where K is the number of samples. The Chi-square probability can be found in appendix tables published in Reference 11. The Kruskal-Wallis One-Way Analysis of Variance by Ranks is used when assumptions for the parametric Analysis of Variance are not suitable for the data, or when the level of data is less than interval measures.

Appendix 1: Common Statistical Tests and their Use

Type of Test	*Use*
Correlation	These tests look for an association between variables
Pearson correlation	Tests for the strength of the association between two continuous variables
Spearman correlation	Tests for the strength of the association between two ordinal variables (does not rely on the assumption of normal distributed data)
Chi-square	Tests for the strength of the association between two categorical variables
Comparison of Means: look for the difference between the means of variables	
Paired T-test	Tests for difference between two related variables
Independent T-test	Tests for difference between two independent variables
ANOVA	Tests the difference between group means after any other variance in the outcome variable is accounted for
Regression: assess if change in one variable predicts change in another variable	
Simple regression	Tests how change in the predictor variable predicts the level of change in the outcome variable
Multiple regression	Tests how change in the combination of two or more predictor variables predict the level of change in the outcome variable
Non-parametric: are used when the data does not meet assumptions required for parametric tests	
Wilcoxon rank-sum test	Tests for difference between two independent variables - takes into account magnitude and direction of difference
Wilcoxon sign-rank test	Tests for difference between two related variables - takes into account magnitude and direction of difference
Sign test	Tests if two related variables are different – ignores magnitude of change, only takes into account direction

Appendix 2: Parametric and Nonparametric Equivalencies

The table below outlines some common research designs and their appropriate parametric and nonparametric equivalents.

Analysis Type	Parametric Test	Nonparametric Test
Compare means of two independent groups	Independent sample t-test	Wilcoxon rank-sum test
Compare means of the same group at two time points	Paired sample t-test	Wilcoxon signed-rank test
Compare means between three or more independent groups	Analysis of variance (ANOVA)	Kruskal-Wallis test
Estimate degree of association	Pearson's Product Moment correlation	Spearman's rank correlation

Appendix 3: Hypothesis Tests Hierarchy

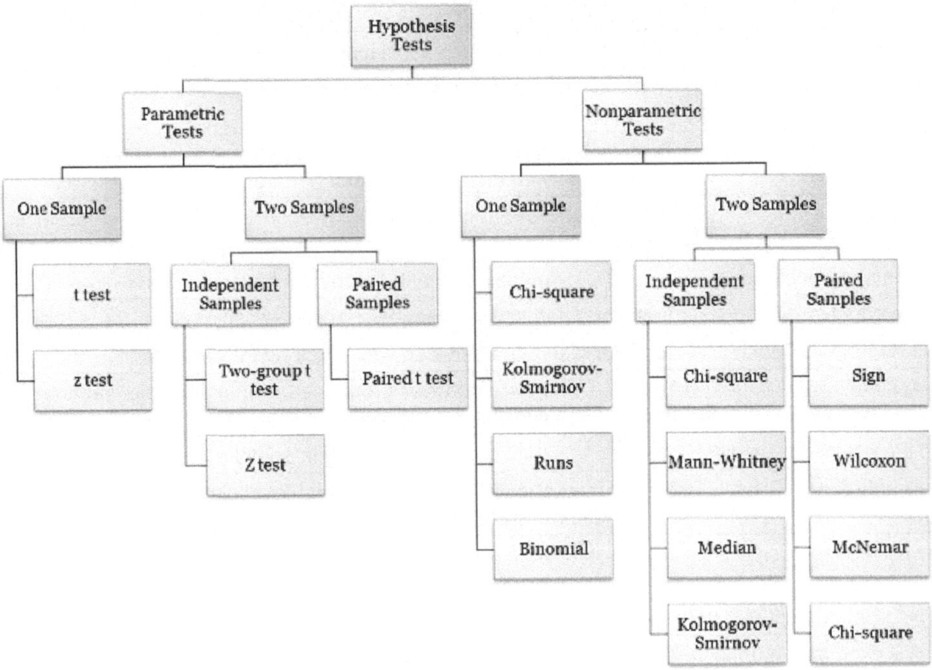

REFERENCES

Agbonifoh, B. A. and Yomere, G. O. (1999).Research Methodology in the Social Sciences and Education, Benin City, Centerpiece Publishers.American Statistician. Vol. 61, No. 1: 47 – 55.

Aweriale, P. E. and Dibua, V. A. (1997).Research methods and Statistics, Benin City, Olive Publishers.

Baccheti, F. (2002). Peer Review of Statistics in Medical Research: The Other Problem. British Medical

Cohen, J. (1998). Statistical Power Analysis for the Behavioural Science, New Jersey.Hillsdale Publishers.

Conover, W.J. (1980). Practical Nonparametric Statistics, New York: Wiley and Sons.

Currier DP. Elements of research in physical therapy. 3rd ed. Baltimore: Williams and Wilkins, 1990.

Domholdt B. Physical therapy research: principles and applications. Philadelphia: WB. Saunders Co., 1993.

Egbule, J. F. and Okobia, D. O, (2001).Research Methods in Education for Colleges and Universities, Agbor, Kmensuo Publishers.

Ferguson GA. Statistical analyses in psychology and education. 5th ed. New York: McGraw-Hill Book Co., 1981.

Fletcher, K. E, French, C. T, Corapi, K. M, Irowins, R. S. and Norman, G. R. (2010). Prospective measures

Freund JE. Modern elementary statistics, 5th ed. Englewood Cliffs, N.J.: Prentice-Hall Inc., 1979.

Gaito, J. (1980). Measurement Scale and Statistics: Resurgence of an Old Misconception, Psychological Bulletin.87, 564 - 567.

Glass G~ Hopkins KD. Statistical methods in education and psychology. 2nd ed. Englewood Cliffs, N.J.: Prentice-Hall Inc., 1984.

Jameison, S. (2004). Likert Scales: How to Abuse Them. Medical Education, 38, 1217 – 1218Kim, M. (2006). Statistical Methods in Arthritis and Rheumatism, Arthritis and Rheumatism Vol. 54, No. 12:Journal, 234, 1271 – 1273.

Kress, G. (1982). Marketing Research, U.S.A, Prentice Hall Publishers.

Kuzon, W. M., Urbanchek, M. G. and McCabe, S. (1996). The Seven Deadly Sins of Statistical Analysis.Annals of Plastic Surgery. 37, 265 – 272.

Lunsford BR. Statistics: screening and data summary. JPO 1993; 5:4:125-30.

Lunsford TR, Lunsford BR. The research sample, part III: sample size. JPO 1995; 7:4:137-41.

McCollough C. Introduction to statistical analyses: a semi-programmed approach. New York: McGraw-Hill Book Co., 1974.

Motulsky, H. (1995). Intuitive Biostatistics, New York: Oxford University Press.

Nachmias, C. F. and Nachmias, D. (2009).Research Methods in the Social Sciences, London, Replika Press Ltd.

Nalou, A. A. (2011). Assessing the Statistical Methodologies of Business Research in South Africa.

Nickerson, R. S. (2000). "Null Hypothesis Significance Testing: A Review of an Old and Continuing

Norusis M. The SPSS guide to data analysis for SPSS-X. Chicago: SPSS Inc., 1987.

Olannye, P. A. (2006). Research Methods for Business: A Skill Building Approach. Lagos, Pee Jen Publishers.

Pearson, E. S. (1981). The Test of Significance for the correction coefficient. Journal of the American Statistical Association 27, 128 – 134.

Rosner, B. (2000). Fundamentals of Biostatistics, California: Duxbury Press.

Rousseau, G. G. (1992). Identifying Criteria of Consumer Awareness in a New South Africa, Management Dynamics. Vol. 1, No.: 67 – 86.

Schmudt, M. J. and Svend, H. (2006).Marketing Research.New York, Prentice Hall Publishers.

Schor S. Fundamentals of biostatistics. New York: GP Putnam Sons, 1969.

Siegel 5, Castallan NJ. Nonparametric statistics for the behavioral sciences, 2nd ed. New York: McGraw-Hill Book Co., 1988.

Siegel S. Nonparametric statistics for the behavioral sciences. New York: McGraw-Hill Book Co., 1956.

Stresak, A. A, Zaman, Q. Marinell, G. Pfeifer. Kp. And Ulmer, H. (2007). The Use of Statistics in Medical

Thorndike R. Correlational Procedures for Research. New York: Gardner Press Inc., 1976.

Tukey JW Quick and dirty methods in statistics: 2. simple analyses for standard designs. In: Quality control conference papers. New York: American Society of Quality Control, 1951.

Walsh, J.E. (1962) Handbook of Nonparametric Statistics, New York: D.V. Nostrand.

Williamsen EW. Statistical reasoning. San Francisco: WH. Freeman and Co., 1974.

— *Chapter 3* —

Sampling Methods

In statistics, sampling is concerned with the selection of a subset of individuals from within a statistical population to estimate characteristics of the whole population. Two advantages of sampling are that the cost is lower and data collection is faster than measuring the entire population. Each observation measures one or more properties (such as weight, location, colour) of observable bodies distinguished as independent objects or individuals. In survey sampling, weights can be applied to the data to adjust for the sample design, particularly stratified sampling. Results from probability theory and statistical theory are employed to guide the practice. In business and medical research, sampling is widely used for gathering information about a population. Acceptance sampling is used to determine if a production lot of material meets the governing specifications.

The sampling process comprises several stages:

- Defining the population of concern
- Specifying a sampling frame, a set of items or events possible to measure
- Specifying a sampling method for selecting items or events from the frame
- Determining the sample size
- Implementing the sampling plan
- Sampling and data collecting

3.1 Basic Concepts of Sampling

1. Population

A population can be defined as including all people or items with the characteristic one wish to understand. Because there is very rarely enough time or money to gather information from everyone or everything in a population, the goal becomes finding a representative sample (or subset) of that population.

Sometimes what defines a population is obvious. For example, a manufacturer needs to decide whether a batch of material from production is of high enough quality to be released to the customer, or should be sentenced for scrap or rework due to poor quality. In this case, the batch is the population.

Although the population of interest often consists of physical objects, sometimes we need to sample over time, space, or some combination of these dimensions. For instance, an investigation of supermarket staffing could examine checkout line length at various times, or a study on endangered penguins might aim to understand their usage of various hunting grounds over time. For the time dimension, the focus may be on periods or discrete occasions.

2. Sampling Frame

We seek a sampling frame which has the property that we can identify every single element and include any in our sample. The most straightforward type of frame is a list of elements of the population (preferably the entire population) with appropriate contact information. For example, in an opinion poll, possible sampling frames include an electoral register and a telephone directory.

A probability sample is a sample in which every unit in the population has a chance (greater than zero) of being selected in the sample, and this probability can be accurately determined. The combination of these traits makes it possible to produce unbiased estimates of population totals, by weighting sampled units according to their probability of selection.

Example: We want to estimate the total income of adults living in a given street. We visit each household in that street, identify all adults living there, and randomly select one adult from each household. (For example, we can allocate each person a random number, generated from a uniform distribution between 0 and 1, and select the person with the highest number in each household). We then interview the selected person and find their income.

People living on their own are certain to be selected, so we simply add their income to our estimate of the total. But a person living in a household of two adults has only a one-in-two chance of selection. To reflect this, when we come to such a household, we would count the selected person's income twice towards the total. (The person who is selected from that household can be loosely viewed as also representing the person who isn't selected.) In the above example, not everybody has the same probability of selection; what makes it a probability sample is the fact that each person's probability is known. When every element in the population *does* have the same probability of selection, this is known as an 'equal probability of selection' (EPS) design. Such designs are also referred to as 'self-weighting' because all sampled units are given the same weight.

3. Sampling Errors and Biases

Sampling errors and biases are induced by the sample design. They include:

1. **Selection bias**: When the true selection probabilities differ from those assumed in calculating the results.
2. **Random sampling error**: Random variation in the results due to the elements in the sample being selected at random.

4. Non-sampling Error

Non-sampling errors are other errors which can impact the final survey estimates, caused by problems in data collection, processing, or sample design. They include:

1. **Over-coverage**: Inclusion of data from outside of the population.
2. **Under-coverage**: Sampling frame does not include elements in the population.
3. **Measurement error**: *e.g.* when respondents misunderstand a question, or find it difficult to answer.
4. **Processing error**: Mistakes in data coding.
5. **Non-response or Participation bias**: Failure to obtain complete data from all selected individuals.

After sampling, a review should be held of the exact process followed in sampling, rather than that intended, in order to study any effects that any divergences might have on subsequent analysis.

5. Non-Response Error

A particular problem is that of *non-response*. Two major types of non-response exist: unit non-response (referring to lack of completion of any part of the survey) and item non-response (submission or participation in survey but failing to complete one or more components/questions of the survey). In survey sampling, many of the individuals identified as part of the sample may be unwilling to participate, not have the time to participate (opportunity cost), or survey administrators may not have been able to contact them. In this case, there is a risk of differences, between respondents and non-respondents, leading to biased estimates of population parameters.

This is often addressed by improving survey design, offering incentives, and conducting follow-up studies which make a repeated attempt to contact the unresponsive and to characterize their similarities and differences with the rest of the frame. The effects can also be mitigated by weighting the data when population benchmarks are available or by imputing data based on answers to other questions. Non response is particularly a problem in internet sampling. Reasons for this problem include improperly designed surveys, over-surveying (or survey fatigue), and the fact that potential participants hold multiple e-mail addresses, which they don't use anymore or don't check regularly.

6. Re-sampling

In statistics, re-sampling is any of a variety of methods for doing one of the following:

1. Estimating the precision of sample statistics (medians, variances, percentiles) by using subsets of available data (jackknifing) or drawing randomly with replacement from a set of data points (bootstrapping)
2. Exchanging labels on data points when performing significance tests (permutation tests, also called exact tests, randomization tests, or re-randomization tests)
3. Validating models by using random subsets (bootstrapping, cross validation)

3.2 Types of Sampling Methods

The sampling techniques are categorized into two groups:

- ☆ Random sampling (Probability sampling)
- ☆ Non-random sampling (Non-probability sampling)

3.2.1 Probability Sampling

Probability sampling is one in which every unit of the population has certain chance of being selected in the sample. It offers a high degree of representative. Probability sampling includes: Simple Random Sampling, Systematic Sampling, Stratified Sampling, Probability Proportional to Size Sampling, and Cluster or Multistage Sampling. These various ways of probability sampling have two things in common:

1. Every element has a known nonzero probability of being sampled and
2. Involves random selection at some point.

1. Simple Random Sampling

In a simple random sample (SRS) of a given size, all such subsets of the frame are given an equal probability. Furthermore, any given *pair* of elements has the same chance of selection as any other such pair (and similarly for triples, and so on). This minimizes bias and simplifies analysis of results. In particular, the variance between individual results within the sample is a good indicator of variance in the overall population, which makes it relatively easy to estimate the accuracy of results.

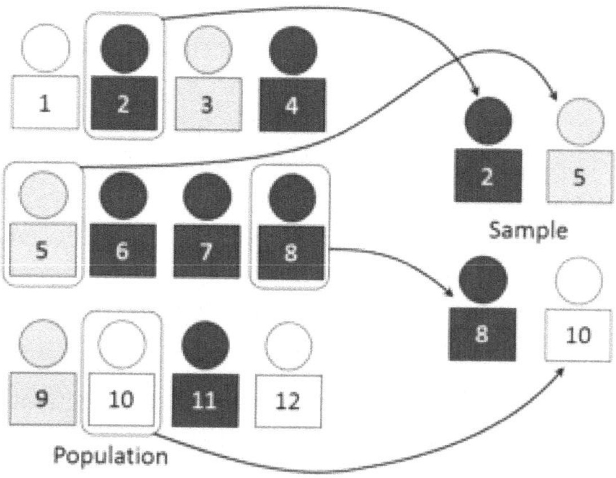

A visual representation of selecting a simple random sample is described in the above diagram. SRS can be vulnerable to sampling error because the randomness of the selection may result in a sample that doesn't reflect the makeup of the population. For instance, a simple random sample of ten people from a given country will *on average* produce five men and five women, but any given trial is likely to over represent one sex and under represent the other. Systematic and stratified techniques attempt to overcome this problem by "using information about the population" to choose a more "representative" sample.

SRS may also be cumbersome and tedious when sampling from an unusually large target population. In some cases, investigators are interested in "research questions specific" to subgroups of the population. For example, researchers might be interested in examining whether cognitive ability as a predictor of job performance is equally applicable across racial groups. SRS cannot accommodate the needs of researchers in this situation because it does not provide subsamples of the population. "Stratified sampling" addresses this weakness of SRS.

2. Systematic Sampling

A visual representation of selecting a random sample using the systematic sampling technique

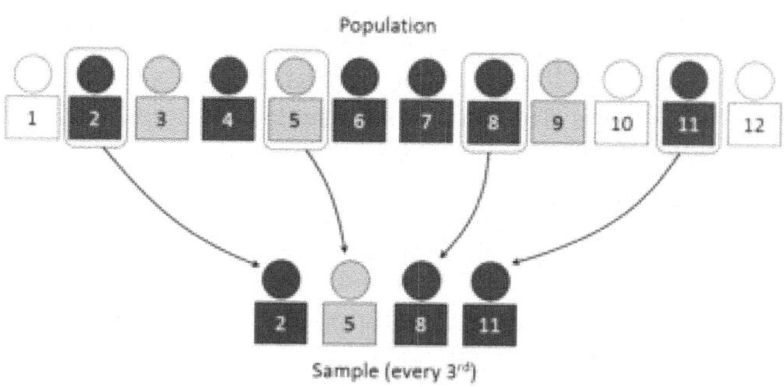

Systematic sampling (also known as interval sampling) relies on arranging the study population according to some ordering scheme and then selecting elements at regular intervals through that ordered list. Systematic sampling involves a random start and then proceeds with the selection of every *k*th element from then onwards. In this case, *k*=(population size/sample size). It is important that the starting point is not automatically the first in the list, but is instead randomly chosen from within the first to the *k*th element in the list. A simple example would be to select every 10th name from the telephone directory (an 'every 10th' sample, also referred to as 'sampling with a skip of 10').

As long as the starting point is randomized, systematic sampling is a type of probability sampling. It is easy to implement and the stratification induced can make it efficient, *if* the variable by which the list is ordered is correlated with the variable of interest. 'Every 10th' sampling is especially useful for efficient sampling from databases.

For example, suppose we wish to sample people from a long street that starts in a poor area (house No. 1) and ends in an expensive district (house No. 1000). A simple random selection of addresses from this street could easily end up with too many from the high end and too few from the low end (or vice versa), leading to an unrepresentative sample. Selecting (*e.g.*) every 10th street number along the street ensures that the sample is spread evenly along the length of the street, representing all of these districts. (Note that if we always start at house #1 and end at #991, the sample is slightly biased towards the low end; by randomly selecting the start between #1 and #10, this bias is eliminated.

However, systematic sampling is especially vulnerable to periodicities in the list. If periodicity is present and the period is a multiple or factor of the interval used, the sample is especially likely to be *un*representative of the overall population, making the scheme less accurate than simple random sampling.

For example, consider a street where the odd-numbered houses are all on the north (expensive) side of the road, and the even-numbered houses are all on the south (cheap) side. Under the sampling scheme given above, it is impossible to get a representative sample; either the houses sampled will *all* be from the odd-numbered, expensive side, or they will *all*be from the even-numbered, cheap side, unless the researcher has previous knowledge of this bias and avoids it by a using a skip which ensures jumping between the two sides (any odd-numbered skip).

Another drawback of systematic sampling is that even in scenarios where it is more accurate than SRS, its theoretical properties make it difficult to *quantify* that accuracy. (In the two examples of systematic sampling that are given above, much of the potential sampling error is due to variation between neighbouring houses – but because this method never selects two neighbouring houses, the sample will not give us any information on that variation.)

As described above, systematic sampling is an EPS method, because all elements have the same probability of selection (in the example given, one in ten). It is *not* 'simple random sampling' because different subsets of the same size have different selection probabilities–*e.g.* the set {4,14,24,.,994} has a one-in-ten probability of selection, but the set {4,13,24,34,.} has zero probability of selection.

Systematic sampling can also be adapted to a non-EPS approach; for an example, see discussion of PPS samples below.

3. Stratified Sampling

When the population embraces a number of distinct categories, the frame can be organized by these categories into separate "strata." Each stratum is then sampled as an independent sub-population, out of which individual elements can be randomly selected.

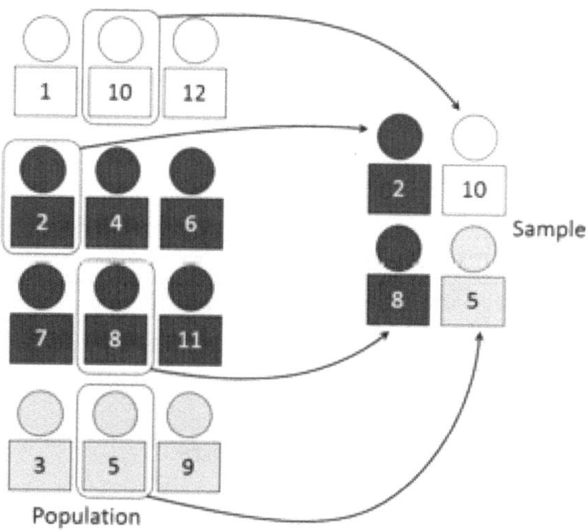

A visual representation of selecting a random sample using the stratified sampling technique is described in the above diagram. There are several potential benefits to stratified sampling.

First, dividing the population into distinct, independent strata can enable researchers to draw inferences about specific subgroups that may be lost in a more generalized random sample. Second, utilizing a stratified sampling method can lead to more efficient statistical estimates (provided that strata are selected based upon relevance to the criterion in question, instead of availability of the samples). Even if a stratified sampling approach does not lead to increased statistical efficiency, such a tactic will not result in less efficiency than would simple random sampling, provided that each stratum is proportional to the group's size in the population. Third, it is sometimes the case that data are more readily available for individual, pre-existing strata within a population than for the overall population; in such cases, using a stratified sampling approach may be more convenient than aggregating data across groups (though this may potentially be at odds with the previously noted importance of utilizing criterion-relevant strata).

Finally, since each stratum is treated as an independent population, different sampling approaches can be applied to different strata, potentially enabling researchers to use the approach best suited (or most cost-effective) for each identified subgroup within the population.

There are, however, some potential drawbacks to using stratified sampling. First, identifying strata and implementing such an approach can increase the cost and complexity of sample selection, as well as leading to increased complexity of population estimates. Second, when examining multiple criteria, stratifying variables may be related to some, but not to others, further complicating the design, and potentially reducing the utility of the strata. Finally, in some cases (such as designs with a large number of strata, or those with a specified minimum sample size per group), stratified sampling can potentially require a larger sample than would other methods (although in most cases, the required sample size would be no larger than would be required for simple random sampling).

A stratified sampling approach is most effective when three conditions are met

1. Variability within strata are minimized
2. Variability between strata are maximized
3. The variables upon which the population is stratified are strongly correlated with the desired dependent variable.

Advantages over other Sampling Methods

1. Focuses on important subpopulations and ignores irrelevant ones.
2. Allows use of different sampling techniques for different subpopulations.
3. Improves the accuracy/efficiency of estimation.
4. Permits greater balancing of statistical power of tests of differences between strata by sampling equal numbers from strata varying widely in size.

Disadvantages

1. Requires selection of relevant stratification variables which can be difficult.
2. Is not useful when there are no homogeneous subgroups.
3. Can be expensive to implement.

Post Stratification

Stratification is sometimes introduced after the sampling phase in a process called "post stratification". This approach is typically implemented due to a lack of prior knowledge of an appropriate stratifying variable or when the experimenter lacks the necessary information to create a stratifying variable during the sampling phase. Although the method is susceptible to the pitfalls of post hoc approaches, it can provide several benefits in the right situation. Implementation usually follows a simple random sample. In addition to allowing for stratification on an ancillary variable, post stratification can be used to implement weighting, which can improve the precision of a sample's estimates.

Oversampling

Choice-based sampling is one of the stratified sampling strategies. In choice-based sampling, the data are stratified on the target and a sample is taken from each stratum so that the rare target class will be more represented in the sample. The model is then built on this biased sample. The effects of the input variables on the target are often estimated with more precision with the choice-based sample even when a smaller overall sample size is taken, compared to a random sample. The results usually must be adjusted to correct for the oversampling.

4. Probability-Proportional-to-Size Sampling

In some cases the sample designer has access to an "auxiliary variable" or "size measure", believed to be correlated to the variable of interest, for each element in the population. These data can be used to improve accuracy in sample design. One option is to use the auxiliary variable as a basis for stratification, as discussed above.

Another option is probability proportional to size ('PPS') sampling, in which the selection probability for each element is set to be proportional to its size measure, up to a maximum of 1. In a simple PPS design, these selection probabilities can then be used as the basis for Poisson sampling. However, this has the drawback of variable sample size, and different portions of the population may still be over- or under-represented due to chance variation in selections.

Systematic sampling theory can be used to create a probability proportionate to size sample. This is done by treating each count within the size variable as a single sampling unit. Samples are then identified by selecting at even intervals among these counts within the size variable. This method is sometimes called PPS-sequential or monetary unit sampling in the case of audits or forensic sampling.

Example: Suppose we have six schools with populations of 150, 180, 200, 220, 260, and 490 students respectively (total 1500 students), and we want to use student population as the basis for a PPS sample of size three. To do this, we could allocate the first school numbers 1 to 150, the second school 151 to 330 (= 150 + 180), the third school 331 to 530, and so on to the last school (1011 to 1500). We then generate a random start between 1 and 500 (equal to 1500/3) and count through the school populations by multiples of 500. If our random start was 137, we would select the schools which have been allocated numbers 137, 637, and 1137, *i.e.* the first, fourth, and sixth schools.

The PPS approach can improve accuracy for a given sample size by concentrating sample on large elements that have the greatest impact on population estimates. PPS sampling is commonly used for surveys of businesses, where element size varies greatly and auxiliary information is often available—for instance, a survey attempting to measure the number of guest-nights spent in hotels might use each hotel's number of rooms as an auxiliary variable. In some cases, an older measurement of the variable of interest can be used as an auxiliary variable when attempting to produce more current estimates.

5. Cluster Sampling

Sometimes it is more cost-effective to select respondents in groups ('clusters'). Sampling is often clustered by geography, or by time periods. (Nearly all samples are in some sense 'clustered' in time – although this is rarely taken into account in the analysis.)

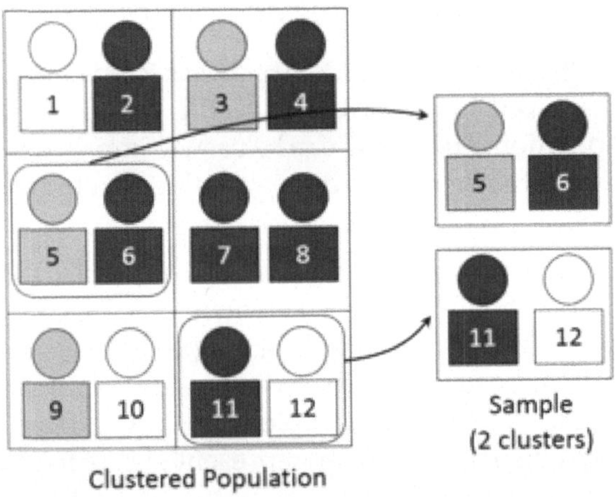

A visual representation of selecting a random sample using the cluster sampling technique is described in the above diagram. For instance, if surveying households within a city, we might choose to select 100 city blocks and then interview every household within the selected blocks.

Clustering can reduce travel and administrative costs. In the example above, an interviewer can make a single trip to visit several households in one block, rather than having to drive to a different block for each household. It also means that one does not need a sampling frame listing all elements in the target population. Instead, clusters can be chosen from a cluster-level frame, with an element-level frame created only for the selected clusters. In the example above, the sample only requires a block-level city map for initial selections, and then a household-level map of the 100 selected blocks, rather than a household-level map of the whole city.

Cluster sampling (also known as clustered sampling) generally increases the variability of sample estimates above that of simple random sampling, depending on how the clusters differ between one another as compared to the within-cluster variation. For this reason, cluster sampling requires a larger sample than SRS to achieve the same level of accuracy – but cost savings from clustering might still make this a cheaper option.

Cluster sampling is commonly implemented as multistage sampling. This is a complex form of cluster sampling in which two or more levels of units are embedded one in the other. The first stage consists of constructing the clusters that will be used to sample from. In the second stage, a sample of primary units is randomly selected from each cluster (rather than using all units contained in all selected clusters). In following stages, in each of those selected clusters, additional samples of units are selected, and so on. All ultimate units (individuals, for instance) selected at the last step of this procedure are then surveyed. This technique, thus, is essentially the process of taking random subsamples of preceding random samples.

Multistage sampling can substantially reduce sampling costs, where the complete population list would need to be constructed (before other sampling methods could be applied). By eliminating the work involved in describing clusters that are not selected, multistage sampling can reduce the large costs associated with traditional cluster sampling. However, each sample may not be a full representative of the whole population.

3.1.2 Non-probability Sampling

Non-probability sampling is any sampling method where some elements of the population have *no* chance of selection (these are sometimes referred to as 'out of coverage'/'undercovered'), or where the probability of selection can't be accurately determined. It involves the selection of elements based on assumptions regarding the population of interest, which forms the criteria for selection. Hence, because the selection of elements is non-random, non-probability sampling does not allow the estimation of sampling errors. These conditions give rise to exclusion bias, placing limits on how much information a sample can provide about the population. Information about the relationship between sample and population is limited, making it difficult to extrapolate from the sample to the population.

Example: We visit every household in a given street, and interview the first person to answer the door. In any household with more than one occupant, this is a non-probability sample, because some people are more likely to answer the door (*e.g.* an unemployed person who spends most of their time at home is more likely to answer than an employed housemate who might be at work when the interviewer calls) and it's not practical to calculate these probabilities. Non-probability sampling methods include convenience sampling, quota sampling and purposive sampling.

1 Quota Sampling

In quota sampling, the population is first segmented into mutually exclusive sub-groups, just as in stratified sampling. Then judgment is used to select the subjects or units from each segment based on a specified proportion. For example, an interviewer may be told to sample 200 females and 300 males between the age of 45 and 60.

It is this second step which makes the technique one of non-probability sampling. In quota sampling the selection of the sample is non-random. For example, interviewers might be tempted to interview those who look most helpful. The problem is that these samples may be biased because not everyone gets a chance of selection. This random element is its greatest weakness and quota versus probability has been a matter of controversy for several years.

2. Mini-max Sampling

In imbalanced datasets, where the sampling ratio does not follow the population statistics, one can resample the dataset in a conservative manner called mini-max sampling. The mini-max sampling has its origin in Anderson mini-max ratio whose value is proved to be 0.5: in a binary classification, the class-sample sizes should be chosen equally. This ratio can be proved to be mini-max ratio only under the assumption of LDA classifier with Gaussian distributions. The notion of mini-max sampling is recently developed for a general class of classification rules, called class-wise smart classifiers. In this case, the sampling ratio of classes is selected so that the worst case classifier error over all the possible population statistics for class prior probabilities, would be the

3. Accidental Sampling

Accidental sampling (sometimes known as grab, convenience or opportunity sampling) is a type of non-probability sampling which involves the sample being drawn from that part of the population which is close to hand. That is, a population is selected because it is readily available and convenient. It may be through meeting the person or including a person in the sample when one meets them or chosen by finding them through technological means such as the internet or through phone. The researcher using such a sample cannot scientifically make generalizations about the total population from this sample because it would not be representative enough. For example, if the interviewer were to conduct such a survey at a shopping centre early in the morning on a given day, the people that he/she could interview would be limited to those given there at that given time, which would not represent the views of other members of society in such an area, if the survey were to be conducted at different times of day and several times per week. This type of sampling is most useful for pilot testing. Several important considerations for researchers using convenience samples include:

1. Are there controls within the research design or experiment which can serve to lessen the impact of a non-random convenience sample, thereby ensuring the results will be more representative of the population?
2. Is there good reason to believe that a particular convenience sample would or should respond or behave differently than a random sample from the same population?
3. Is the question being asked by the research one that can adequately be answered using a convenience sample?

4. Voluntary Sampling

The voluntary sampling method is a type of non-probability sampling. A voluntary sample is made up of people who self-select into the survey. Often, these subjects have a strong interest in the main topic of the survey. Volunteers may be invited through advertisements on Social Media Sites. This method is suitable for a research which can be done through filling a questionnaire. The target population for advertisements can be selected by characteristics like demography, age, gender, income, occupation, education level or interests using advertising tools provided by the social media sites. The advertisement may include a message about the research and will link to a web survey. After voluntary following the link and submitting the web based questionnaire, the respondent will be included in the sample population. This method can reach a global population and limited by the advertisement budget. This method may permit volunteers outside the reference population to volunteer and get included in the sample. It is difficult to make generalizations about the total population from this sample because it would not be representative enough.

5. Line-intercept Sampling

Line-intercept sampling is a method of sampling elements in a region whereby an element is sampled if a chosen line segment, called a "transect", intersects the element.

6. Panel Sampling

Panel sampling is the method of first selecting a group of participants through a random sampling method and then asking that group for (potentially the same) information several times over a period of time. Therefore, each participant is interviewed at two or more time points; each period of data collection is called a "wave". The method was developed by sociologist Paul Lazarsfeld in 1938 as a means of studying political campaigns. This longitudinal sampling-method allows estimates of changes in the population, for example with regard to chronic illness to job stress to weekly food expenditures. Panel sampling can also be used to inform researchers about within-person health changes due to age or to help explain changes in continuous dependent variables such as spousal interaction. There have been several proposed methods of analyzing panel data, including MANOVA, growth curves, and structural equation modeling with lagged effects.

7. Snowball Sampling

In social science research, Snowball sampling involves finding a small group of initial respondents and using them to recruit more respondents. It is particularly useful in cases where the population is hidden or difficult to enumerate. Some variants of snowball sampling, such as respondent driven sampling, allow calculation of selection probabilities and are probability sampling methods under certain conditions.

REFERENCES

[1. *"Examples of sampling methods"* (*PDF*). Cohen, 1988

2. *"Voluntary Sampling Method"*. Lazarsfeld, P., and Fiske, M. (1938). The" panel" as a new tool for measuring opinion. The Public Opinion Quarterly, 2(4), 596–612.

3. *Anderson, Theodore (1951). "Classification by multivariate analysis". Psychometrika.* ***16*** *(1): 31–50. doi:10.1007/bf02313425.*

4. Berinsky, A. J. (2008). Survey non-response. In W. Donsbach and M. W. Traugott (Eds.), The SAGE handbook of public opinion research (pp. 309–321). Thousand Oaks, CA: Sage Publications.

5. Chambers, R L, and Skinner, C J (editors) (2003), *Analysis of Survey Data*, Wiley, ISBN 0-471-89987-9

6. David S. Moore and George P. McCabe. *"Introduction to the Practice of Statistics"*.

7. *DeepanPalguna, Vikas Joshi, VenkatesanChakaravarthy, Ravi Kothari and L. V. Subramaniam (2015). Analysis of Sampling Algorithms for Twitter. International Joint Conference on Artificial Intelligence.*

8. Deming, W. Edwards (1975) On probability as a basis for action, *The American Statistician*, 29(4), pp146–152.

9. Dillman, D.A., Smyth, J.D., and Christian, L. M. (2009). Internet, mail, and mixed-mode surveys: The tailored design method. San Francisco: Jossey-Bass.

10. Gy, P (1992) *Sampling of Heterogeneous and Dynamic Material Systems: Theories of Heterogeneity, Sampling and Homogenizing* **J** Salant, Priscilla, I. Dillman, and A. Don. *How to conduct your own survey*. No. 300.723 S3. 1994.

11. Dillman, D. A., Eltinge, J. L., Groves, R. M., and Little, R. J. A. (2002). Survey nonresponse in design, data collection, and analysis. In R. M. Groves, 12. D. A. Dillman, J. L. Eltinge, and R. J. A. Little (Eds.), Survey nonresponse (pp. 3–26). New York: John Wiley and Sons.

13. Korn, E.L., and Graubard, B.I. (1999) *Analysis of Health Surveys*, Wiley, ISBN 0-471-13773-1

14. *Lance, P. and Hattori, A. (2016). Sampling and Evaluation. Web: MEASURE Evaluation. pp. 6–8; 62–64.*

15. *Lohr, Sharon L. Sampling: Design and analysis.*

16. *Lohr, Sharon L. Sampling: Design and Analysis.Särndal, Carl-Erik, and Swensson, Bengt, and Wretman, Jan. Model Assisted Survey Sampling.*

17. Lucas, Samuel R. (2012). "Beyond the Existence Proof: Ontological Conditions, Epistemological Implications, and In-Depth Interview Research.", *Quality and Quantity*, doi:10.1007/s11135-012-9775-3.

18. Porter, Whitcomb, Weitzer (2004) Multiple surveys of students and survey fatigue. In S. R. Porter (Ed.), Overcoming survey research problems: Vol. 121. New directions for institutional research (pp. 63–74). San Francisco, CA: Jossey Bass.

19. *Robert M. Groves; et al., Survey methodology ISBN 0470465468.*

20. *Särndal, Carl-Erik, and Swensson, Bengt, and Wretman, Jan. Model Assisted Survey Sampling.*

21. *Scheaffer, Richard L., William Mendenhal and R. Lyman Ott. Elementary survey sampling.*

22. *Scott, A.J.; Wild, C.J. (1986). "Fitting logistic models under case-control or choice-based sampling". Journal of the Royal Statistical Society, Series B.* ***48****: 170–182. JSTOR 2345712.*

23. *ShahrokhEsfahani, Mohammad; Dougherty, Edward (2014). "Effect of separate sampling on classification accuracy". Bioinformatics.* ***30*** *(2): 242–25. PMID 24257187. doi:10.1093/bioinformatics/btt662.*

24. *Smith, T. M. F. (1984). "Present Position and Potential Developments: Some Personal Views: Sample surveys". Journal of the Royal Statistical Society, Series A.* ***147*** *(The 150th Anniversary of the Royal Statistical Society, number 2): 208–221. JSTOR 2981677. doi:10.2307/2981677.*

25. *Smith, T. M. F. (1993). "Populations and Selection: Limitations of Statistics (Presidential address)". Journal of the Royal Statistical Society, Series A.* ***156*** *(2): 144–166. JSTOR 2982726. doi:10.2307/2982726.* (Portrait of T. M. F. Smith on page 144)

26. *Smith, T. M. F. (2001). "Biometrika centenary: Sample surveys". Biometrika.* ***88*** *(1): 167–243. doi:10.1093/biomet/88.1.167.*

27. *Smith, T. M. F. (2001). "Biometrika centenary: Sample surveys". In D. M. Titterington and D. R. Cox. Biometrika: One Hundred Years. Oxford University Press. pp. 165–194. ISBN 0-19-850993-6.*

28. Stuart, Alan (1962) *Basic Ideas of Scientific Sampling,* Hafner Publishing Company, New York

29. Vehovar, V., Batagelj, Z., Manfreda, K.L., and Zaletel, M. (2002). Nonresponse in web surveys. In R. M. Groves, D. A. Dillman, J. L. Eltinge, and R. J. A. Little (Eds.), Survey nonresponse (pp. 229–242). New York: John Wiley and Sons.

30. *Whittle, P. (May 1954). "Optimum preventative sampling". Journal of the Operations Research Society of America.* **2** *(2): 197–203. JSTOR 166605. doi:10.1287/opre.2.2.197.*

— *Chapter 4* —

Advanced Statistical Tools for Research Data Analysis

Multivariate analysis is essentially the statistical process of simultaneously analysing multiple independent variables with multiple dependent (outcome or criterion) variables using matrix algebra. During the last two or three decades, multivariate statistical analysis has become increasingly popular. The theory has made great progress, and with the rapid advances in computer technology, routine applications of multivariate statistical methods are implemented in several statistical software packages. The traditional approach to the teaching of multivariate statistical analysis, as exemplified by Anderson (1958), relies heavily on advanced matrix mathematics. On the other hand, Hury and Riedwyl (1988) suggest that it is possible to understand most of the basic ideas underlying multivariate statistical analysis without a mastery of such mathematics, provided that these are conveyed with the help of real data sets. In this chapter we propose a non-mathematical data-driven approach for teaching multivariate statistical methods to students. Despite this, we are mindful of the need for students to know some basic linear algebra and uni-variate statistical concepts. Such basic knowledge provides students with the foundation necessary for the application of the appropriate multivariate statistical procedures and for the interpretation of results

4.1 Basic Research Question

A critical aspect of conducting good research involves the type of design used and the type of statistical analysis. These flow from the basic research question asked.

1. **Degree of relationship between the variables.** Bivariate correlation and regression (simple, multiple, and multivariate/canonical) are used to 1) explain the association between variables or 2) predict the value of one or more criterion variables knowing the value of one or more predictor variables. Path analysis is used estimate direct and indirect causal relationships between variables which are observed (directly measured) [note: structural equation modeling and confirmatory factor analysis are similar but also include latent (or unobserved) variables].
2. **Measure significant differences between group means.** MANOVA (multivariate analysis of variance) is used when group differences are measured on two or more dependent variables that are related to one another in the real world (*e.g.*, depression and anxiety). Like ANOVA, MANOVA controls for Type I errors if two or more ANOVAs are run, and different kinds of MANOVA analyses can be conducted: one-way MANOVA, factorial MANOVA, repeated-measures MANOVA, and MANCOVA (using one or more covariates in controlling variance).
3. **Predicting membership in two or more groups from one or more variables.** Outcome studies involving the classification or prediction of subjects into two groups (*e.g.*, success/failure, yes/no, hire/don't hire, *etc.*) are conducted through logistic regression. When subjects are classified into three or more groups, discriminant analysis is used. Predictive discriminant analysis (PDA) is used to predict group membership, while descriptive discriminant analysis (DDA) is used to explain the best linear combination of dependent variables that maximizes group differences. Logistic analysis can use categorical predictors without difficulty, but in discriminant analysis, categorical predictors pose some problems but they are not insurmountable.
4. **Explaining underlying structure.** Often researchers infer that variables may be clustered together based on their common correlations which implies that they are correlated because they are representing some common underlying trait or factor. Principal components analysis (PCA) and common factor analysis (CFA) reduce the variables into smaller subsets based on shared variance.

4.2 Multiple Techniques

4.2.1 Multiple Regressions

Multiple regressions is the most commonly utilized multivariate technique.

It examines the relationship between a single metric dependent variable and two or more metric independent variables. The technique relies upon determining the linear relationship with the lowest sum of squared variances; therefore, assumptions of normality, linearity, and equal variance are carefully observed. The beta coefficients (weights) are the marginal impacts of each variable, and the size of the weight can be interpreted directly. Multiple regression is often used as a forecasting tool. Goal is to use the linear composite of two or more continuous and/or categorical variables (predictors) to: 1) predict scores on a single continuous variable (criterion), or to 2) explain the nature of the single continuous criterion variable from what is known about the predictor variables. In prediction, the criterion is the main emphasis because decisions are made on its value, but often times, it is difficult to directly measure or obtain a subject's actual score on the criterion; therefore, it is important to estimate or predict one's criterion score based on the value of the predictor scores.

4.2.2 Logistic Regression

Logistic regression is used to predict a categorical (usually dichotomous) variable from a set of predictor variables. With a categorical dependent variable, discriminant function analysis is usually employed if all of the predictors are continuous and nicely distributed; logit analysis is usually employed if all of the predictors are categorical; and logistic regression is often chosen if the predictor variables are a mix of continuous and categorical variables and/or if they are not nicely distributed (logistic regression makes no assumptions about the distributions of the predictor variables). Logistic regression has been especially popular with medical research in which the dependent variable is whether or not a patient has a disease. For a logistic regression, the predicted dependent variable is the estimated probability that a particular subject will be in one of the categories (for example, the probability that Suzie Cue has the disease, given her set of scores on the predictor variables).

In a logistic regression, odds ratios are commonly employed to measure the strength of the partial relationship between one predictor and the dependent variable (in the context of the other predictor variables). It may be helpful to consider a simple univariate odds ratio first. Among the male respondents, 68 approved continuing the research, 47 voted to stop it, yielding odds of 68/47. That is, approval was 1.45 times more likely than non approval. Among female respondents, the odds were 60/140. That is, approval was only.43 times as likely as was non approval. Inverting these odds (odds less than one are difficult for some people to comprehend), among female respondents non approval was 2.33 times as likely as approval.

Because the outcome is dichotomous, the relationship between Y and the Xs is non-linear (the line is S-shaped rather than straight), therefore the goal is not to predict a score on Y but to predict the probability that a subject belongs to one group or another. A probability score of 1 indicates absolute certainty that the individual belongs to the target group, and a probability score of 0 indicates absolute certainty that the person does not belong to the target group. Generally a cutting score of 0.5 is used so that those with probabilities above 0.5 are classified into the target group (Y=1.0), and those below 0.5 are classified into the other group (Y=0.0). A linear composite X score is derived from the simultaneous solution of the data matrix, but becomes the exponent for the natural log *e*: $P/(1-P) = e^{a + b1X1 + b2X2 + \dots bnXn}$.Logistic regression analysis does not use ordinary least squares in weighting the X variables, but uses a more complex process of maximum likelihood estimation. MLE requires a greater sample size than does linear multiple regression.

Interpretation

Interpretation follows four steps.

1. The researcher must determine if the overall model is sufficient. An F test is conducted using the likelihood ratio which is based on $\chi 2$. Essentially the ratio involves the likelihood of the full model compared to the model using only the constant as a predictor.
2. One can examine a *pseudo*-R^2 value, which is not a true proportion of variance accounted for by the model, but is an estimated value.
3. Look at a goodness-of-fit index to determine if the model estimated by MLE fits well with the data at hand. The value is -2 (Log Likelihood) or -2LL. A perfect model would have a value of 0. The goal is to have values approaching 0 as more important predictor variables are entered into the model. The -2LL is evaluated each time predictors are added into the model. Selection procedures are similar to those used in multiple regressions.
4. A classification table is given which indicates the hit rate of the model selected. It is based on a contingency table of True +, True -, False +, and False -. SPSS gives the frequency of the four contingencies plus the overall hit rate (True + plus True -). The final hit rate should be greater than that of chance alone which is represented by the base rate (the ratio of actual +/total sample size). This is often called *incremental validity*. Further analysis of hit rates calculates positive predictive values and negative predictive values.

4.2.3 Canonical Correlation

Canonical correlation is another extension of multiple regression where rather than using a single outcome variable Y, two or more Y variables are predicted by two or more predictor X variables. The purpose is usually not to predict Ys from Xs but to explain the relationship between the X and Y variable sets. The analysis can be run in both directions, Xs predicting Ys and Ys predicting Xs, but the researcher is generally concerned about Xs predicting

Ys where the Y set is the new hypothesized inter-relationship and the X set are the traditional predictors seen in prior research.

In Multiple regression, several X variables (predictors) are independent and only one Y variable (criterion) is dependent. In canonical correlation, there are multiple sets of X variables and multiple sets of Y variables.

Because canonical correlation involves multiple Xs and multiple Ys, shared variance exists along two or more dimensions or geometric axes (with one outcome variable in multiple regression, there is only one dimension or axis). In fact, the number of dimensions or axes is equal to the number of variables in the smaller of the two sets. For example, if the data had three X variables and four Y variables, then the number of dimensions would be three. If the data had two X variables and two Y variables, the number of dimensions would be two. Each dimension will be represented by two linear composites, one for the X set of variables and one for the Y set. The correlation between the two composites is called the *canonical correlation*, RC. It is analogous to the multiple R in multiple regression. The RC2, also labeled as the *eigen value*, indicates the proportion of variance shared between the two composites. Like multiple regression where the predictor variable with the greatest degree of shared variance with Y is entered into the model first, the two sets of X and Y variables with the greatest degree of shared variance are entered into the canonical correlation model first. This canonical correlation of the X and Y composite variables is called the 1st *canonical* function. After this first canonical function extracts its proportion of shared variance then a second canonical correlation is computed from the remaining variance. Because the second function does not include any of the variance of the first function, the two sets are totally uncorrelated (orthogonal or independent). Each successive canonical function will be uncorrelated with all previous canonical functions. It is important to also note that because each successive canonical function has less and less remaining variance in which to operate, each successive RC gets smaller than the previous canonical correlations.

Interpretation

Interpretation involves five steps:

1. The overall significance of the model (null: explained variance = 0) is tested by Wilk's lambda. If the model is significant (explained variance > 0), then proceed.
2. Determine which canonical functions are important enough to keep. Two indices are considered simultaneously. A $\chi 2$ based on Wilks lambda is calculated for the total set of canonical functions with the null being that all of the RC = 0. If the chi-square is significant, then one can conclude that at least the first canonical correlation is significantly greater than 0. This is because the first set has the greatest RC value. The second chi-square tests the null that all of the remaining RC = 0. At the point the chi-square becomes non-significant, the remaining

chi-squares will also be non-significant. As mentioned in previous sections, statistical significance is heavily influenced by sample size, so one must also look at the RC2 value of each canonical function. A general rule given by Pedhazur (1997) is to keep functions with RC2 > 10 per cent.

3. As in other regression analyses, regression coefficients will be calculated for each variable in the X and Y set. These are named *canonical weights or coefficients*. These are standardized and their values will be recalculated for each successive canonical function. These can be interpreted as to their respective contribution but have the same interpretation confound seen in other regression coefficients.
4. Structure coefficients are more widely used for determining the importance of each variable. The structure coefficient is the correlation between a variable and its respective linear composite. For example, the structure coefficient for X1 is the correlation of X1 with the X linear composite. The squared structure coefficient is the amount of the linear composite variance that is explained by X1. Most researchers construct a table of structure coefficients much like the table used for factor analysis. The columns represent the canonical functions retained and the rows represent each variable. The cell data represent the structure coefficients. Like factor analysis, the researcher then underscores the structure coefficients with values greater than.45 indicating that these have significant contribution to the variance of the canonical function. Furthermore, the variables which are retained provide a description of the dimension represented in that particular canonical function. These dimensions can even be named as factors are in factor analysis.
5. As discussed, the RC2 represents only the shared variance within a specific canonical function and does not represent the proportion explained in the total X and Y variance. A *redundancy index* is calculated for each canonical function to provide an estimate of the variance explained for the full model. A separate redundancy is calculated for the X set and the Y set. It is the average structure coefficient multiplied by the canonical correlation.

4.2.4 Factor Analysis

The term factor analysis refers to a set of analytical techniques designed to reduce data into smaller, meaningful groups based upon their inter-correlations or shared variance. The assumption is that those items or variables that are correlated must be measuring a similar factor or trait or construct. In the case where only a few variables are used, the researcher may be able to determine groupings by simply observing the content of each variable; however, for large data sets and/or more ambiguous items, this task would be formidable. Factor Analysis reduces the information in a model by reducing the dimensions of the observations. This procedure has multiple purposes. It can be used to simplify the data, for example reducing the number of variables in predictive regression models. If factor analysis is used for these purposes, most often factors are rotated after extraction. Factor analysis has several different rotation methods—some of them ensure that the factors are orthogonal. Then the correlation coefficient between two factors is zero, which eliminates problems of multicollinearity in regression analysis.

Factor analysis is also used in theory testing to verify scale construction and operationalizations. In such a case, the scale is specified upfront and we know that a certain subset of the scale represents an independent dimension within this scale. This form of factor analysis is most often used in structural equation modeling and is referred to as Confirmatory Factor Analysis. For example, we know that the questions pertaining to the big five personality traits cover all five dimensions N, A, O, and I. If we want to build a regression model that predicts the influence of the personality dimensions on an outcome variable, for example anxiety in public places, we would start to model a confirmatory factor analysis of the twenty questionnaire items that load onto five factors and then regress onto an outcome variable.

Factor analysis can also be used to construct indices. The most common way to construct an index is to simply sum up the items in an index. In some contexts, however, some variables might have a greater explanatory power than others. Also sometimes similar questions correlate so much that we can justify dropping one of the questions completely to shorten questionnaires. In such a case, we can use factor analysis to identify the weight each variable should have in the index.

Factor analysis is a data reduction technique that can reduce the number of items by grouping them and by examining the content of the items in each group one can determine the structure or composition of each group thereby giving a better explanation of the data. It is important to note that factor analysis is not used in prediction or explaining the relationship between different sets of variables, nor is it used to determine group differences. The goal is to explain the underlying structure or composition of the data; therefore we are dealing only with one set of variables.

Two types of factor analysis exist. The first, *exploratory factor analysis* (EFA) is used to explore or derive the underlying factor structure of a data matrix often without regard to theory. The purpose of EFA is to determine if underlying factors exist within a data set, and if so, what those factors are. The researcher does not necessarily need to have any expectations or theory beforehand—it can simply be "exploratory." Two types of EFA are commonly used. *Principle components analysis* (PCA) tries to account for all variance among the variables/items and so includes both shared variance and unique/error variance of each variable/item. *Principle factor analysis* (PFA) accounts for only shared variance of each variable/item. Currently, most journals prefer PFA.

The second, *confirmatory factor analysis* (CFA), is used to test a priori theory. The researcher specifies what factors exist and what variables/items constitute each factor and then "orders" these parameters into a data set to determine if indeed the factors and variables/items describe or fit the data. CFA is most commonly conducted through Structural Equation Modeling (SEM). The following discussion is on factor analysis pertaining to EFA.

The main goal of factor analysis is to explain as much variance as possible in a data set by using the smallest number of factors (groupings of variables based on high inter-correlations) and the smallest amount of items or variables within each factor. Inherently one balances explained variance with simplicity. Critical to this technique is that one wants to ensure that the variance left out of the solution is primarily error variance. In EFA, the first factor derived is a linear composite of all variables/items such that it maximizes the amount of total variance explained (or extracted)—no other linear combination will extract as much variance. The proportion of variance extracted is called the *eigen value.* A second factor which is orthogonal to the first (uncorrelated) is then derived from the remaining variance, and its eigen value will be derived. Like the first factor, the second factor will be a linear composite of all variables/items. This process continues until all variance has been extracted, and the number of extractions is equal to the number of original variables/items.

Interpretation

Interpretation involves three steps:

1. The researcher determines the number of factors to keep. Several decision rules can be employed but generally keep those factors with eigen values greater than 1.0. The reason is that any given variable has a variance equal to 1.0 (since variables are standardized, the std. dev. = 1.0, and variance is the std. dev. Squared) which means that eigen values should explain more variance than at least one variable/item. In a satisfactory EFA, the total variance of the retained eigen values should be greater than 70 per cent.
2. Once the number of factors has been determined, the researcher then determines which variables/items "load" on each factor. This is determined by the coefficient of the variable/item. Because each factor is a linear composite of all variables/items, each variable/item will have a different coefficient for each factor. This coefficient is actually a structure coefficient since it is the correlation between the variable/item and the factor; however, it is called the factor loading or the factor structure coefficient. The higher the coefficient, the greater the variable/item's contribution to the factor, and the square of the coefficient is the amount of variance of the factor that is explained by the variable/item. In general, the desired outcome is that each variable/item will have a large loading on only one factor and small loadings on the remaining factors.

Before actually determining the composition of each factor, the axes which represent factor dimensions can be rotated geometrically so that the new set of axes are positioned closer to their respective factor variables/items. These axes are similar to those in a Cartesian coordinate system. For example, in a two-factor model, every variable/item will have a factor loading value for Factor I and for Factor II. Most will have a relatively high loading value (>.5) on one factor and relatively small loading on the other factor (<.5). Each factor is an axis in space and the factor loading of a variable/item is the coordinate value on that axis. If variable X1 has a factor loading of.63 on FI, and.22 on FII, then its coordinate point is (.63,.22). Also, the distance of X1 from the origin can be calculated by the Pythagorean theorem (add.63 squared and.22 squared and take the square root). By rotating the axes, one can increase the value along the FI dimension and decrease the FII dimension value as long as the distance of X1 to the origin remains unchanged. This strengthens the association of X1 with FI while decreasing its association to FII. This makes interpretation of the factor structure (composition) simpler. Usually when rotating the axes, the researcher assumes that the factors are truly orthogonal and so an orthogonal rotation is used, that is the axes remain at 90 degrees to one another. This is easy to visualize in a two-factor model. If F1 is rotated 35 degrees clockwise, FII must also be rotated 35 degrees clockwise. However, sometimes the factors are somewhat correlated and so the rotation that occurs is based upon the correlation between the factors. The cosine of the angle between the axes is equal to the correlation. By the way, this rule holds true when orthogonal rotation is used because the angles are at 90 degrees and the cosine of 90 degrees is 0.

3. With the output arranged in a table where the columns are the factors and the rows are the variables/items, one then highlights or underscores the loadings under each factor that are greater than.50. Once all contributing variables on all factors have been identified, then one must determine the content of each factor and assign an appropriate name for each. This is done by analyzing the content or general theme of the variable/items that are highlighted.

4.2.5 Principal Component Analysis (PCA)

PCA involves a mathematical procedure that transforms a number of (possibly) correlated variables into a (smaller) number of uncorrelated variables called *principal components*. The first principal component accounts for as much of the variability in the data as possible, and each succeeding component accounts for as much of the remaining variability as possible.

Objectives

- ☆ To discover or to reduce the dimensionality of the data set.
- ☆ To identify new meaningful underlying variables.

Traditionally, principal component analysis is performed on the Covariance matrix or on the Correlation matrix. These matrices can be calculated from the data matrix. The covariance matrix contains scaled sums of squares and cross products. A correlation matrix is like a covariance matrix but first the variables, *i.e.* the columns, have been standardized. We will have to standardize the data first if the variances of variables differ much, or if the units of measurement of the variables differ. The mathematical technique used in PCA is called eigen analysis: we solve for the eigen values and eigen vectors of a square symmetric matrix with sums of squares and cross products. The eigenvector associated with the largest eigen value has the same direction as the first principal component. The eigenvector associated with the second largest eigen value determines the direction of the second principal component. The sum of the eigen values equals the trace of the square matrix and the maximum number of eigenvectors equals the number of rows (or columns) of this matrix.

There are two methods to help you to choose the number of components to keep. Both methods are based on relations between the eigen values.

- ☆ Plot the eigen values, If the points on the graph tend to level out (show an "elbow"), these eigen values are usually close enough to zero that they can be ignored.
- ☆ Limit the number of components to that number that accounts for a certain fraction of the total variance.

4.2.6 Multivariate Analysis of Variance (MANOVA)

This technique examines the relationship between several categorical independent variables and two or more metric dependent variables. Whereas analysis of variance (ANOVA) assesses the differences between groups (by using T tests for two means and F tests between three or more means), MANOVA examines the dependence relationship between a set of dependent measures across a set of groups. Typically this analysis is used in experimental design, and usually a hypothesized relationship between dependent measures is used. This technique is slightly different in that the independent variables are categorical and the dependent variable is metric. Sample size is an issue, with 15-20 observations needed per cell. However, too many observations per cell (over 30) and the technique lose its practical significance. Cell sizes should be roughly equal, with the largest cell having less than 1.5 times the observations of the smallest cell. That is because, in this technique, normality of the dependent variables is important. The model fit is determined by examining mean vector equivalents across groups. If there is a significant difference in the means, the null hypothesis can be rejected and treatment differences can be determined. A multivariate analysis of variance (MANOVA) could be used to test this hypothesis. Instead of a univariate *F* value, we would obtain a multivariate *F* value (Wilks' Lambda) based on a comparison of the error variance/covariance matrix and the effect variance/covariance matrix. Although we only mention Wilks' Lambda? Here, there are other statistics that may be used, including Hotelling's trace and Pillai's criterion. MANOVA is useful in experimental situations where at least some of the independent variables are manipulated. It has several advantages over ANOVA.

(a) By measuring several dependent variables in a single experiment, there is a better chance of discovering which factor is truly important.

(b) It can protect against Type I errors that might occur if multiple ANOVA's were conducted independently. Additionally, it can reveal differences not discovered by ANOVA tests.

However, there are several cautions as well.

(a) It is a substantially more complicated design than ANOVA, and therefore there can be some ambiguity about which independent variable affects each dependent variable. Thus, the observer must make many potentially subjective assumptions.

(b) Moreover, one degree of freedom is lost for each dependent variable that is added. The gain of power obtained from decreased SS error may be offset by the loss in these degrees of freedom.

(c) Finally, the dependent variables should be largely uncorrelated. If the dependent variables are highly correlated, there is little advantage in including more than one in the test given the resultant loss in degrees of freedom. Under these circumstances, use of a single ANOVA test would be preferable.

Assumptions

Normal Distribution - The dependent variable should be normally distributed within groups. Overall, the *F* test is robust to non-normality, if the non-normality is caused by skewness rather than by outliers (outliers are values that are very low or very high as compared to the most values in the data set). Tests for outliers should be run before performing a MANOVA, and outliers should be transformed or removed.

Linearity - MANOVA assumes that there are linear relationships among all pairs of dependent variables, all pairs of covariates, and all dependent variable-covariate pairs in each cell. Therefore, when the relationship deviates from linearity, the power of the analysis will be compromised.

Homogeneity of Variances - Homogeneity of variances assumes that the dependent variables exhibit equal levels of variance across the range of predictor variables. Remember that the error variance is computed (SS error) by adding up the sums of squares within each group. If the variances in the two groups are different from each other, then adding the two together is not appropriate, and will not yield an estimate of the common within-group variance.

Homogeneity of Variances and Co-variances - In multivariate designs, with multiple dependent measures, the homogeneity of variances assumption described earlier also applies. However, since there are multiple dependent variables, it is also required that their inter-correlations (co-variances) are homogeneous across the cells of the design. There are various specific tests of this assumption.

Special Cases

Two special cases arise in MANOVA, the inclusion of within-subjects independent variables and unequal sample sizes in cells. Unequal sample sizes - As in ANOVA, when cells in a factorial MANOVA have different sample sizes, the sum of squares for effect plus error does not equal the total sum of squares. This causes tests of main effects and interactions to be correlated. SPSS offers and adjustment for unequal sample sizes in MANOVA.

Within-subjects design - Problems arise if the researcher measures several different dependent variables on different occasions. This situation can be viewed as a within-subject independent variable with as many levels as occasions. Or, it can be viewed as a separate dependent variables for each occasion.

Additional Limitations

Outliers - Like ANOVA, MANOVA is extremely sensitive to outliers. Outliers may produce either a Type I or Type II error and give no indication as to which type of error is occurring in the analysis. There are several programs available to test for univariate and multivariate outliers. Multicollinearity and Singularity - When there is high correlation between dependent variables, one dependent variable becomes a near-linear combination of the other dependent variables. Under such circumstances, it would become statistically redundant and suspect to include both combinations.

4.2.7 Discriminant Function Analysis

Discriminant function analysis is used to determine which variables discriminate between two or more naturally occurring groups. For example, an educational researcher may want to investigate which variables discriminate between high school graduates who decide (1) to go to college, (2) to attend a trade or professional school, or (3) to seek no further training or education. For that purpose the researcher could collect data on numerous variables prior to students' SSLC. After SSLC, most students will naturally fall into one of the three categories. *Discriminant Analysis* could then be used to determine which variable(s) are the best predictors of students' subsequent educational choice.

Stepwise Discriminant Analysis

Probably the most common application of discriminant function analysis is to include many measures in the study, in order to determine the ones that discriminate between groups. For example, an educational researcher interested in predicting high school students' choices for further education would probably include as many measures of personality, achievement motivation, academic performance, *etc.* as possible in order to learn which one(s) offer the best prediction. Put another way, we want to build a "model" of how we can best predict to which group a case belongs. In the following discussion we will use the term "in the model" in order to refer to variables that are included in the prediction of group membership, and we will refer to variables as being "not in the model" if they are not included.

Forward Stepwise Analysis

In stepwise discriminant function analysis, a model of discrimination is built step-by-step. Specifically, at each step all variables are reviewed and evaluated to determine which one will contribute most to the discrimination between groups. That variable will then be included in the model, and the process starts again.

Backward Stepwise Analysis

One can also step backwards; in that case all variables are included in the model and then, at each step, the variable that contributes least to the prediction of group membership is eliminated. Thus, as the result of a successful discriminant function analysis, one would only keep the "important" variables in the model, that is, those variables that contribute the most to the discrimination between groups.

F to Enter, F to Remove

The stepwise procedure is "guided" by the respective *F* to enter and *F* to remove values. The *F* value for a variable indicates its statistical significance in the discrimination between groups, that is, it is a measure of the extent to which a variable makes a unique contribution to the prediction of group membership.

Interpreting a Two-Group Discriminant Function

In the two-group case, discriminant function analysis can also be thought of as (and is analogous to) multiple regression. If we code the two groups in the analysis as *1* and *2*, and use that variable as the dependent variable in a multiple regression analysis, then we would get results that are analogous to those we would obtain via *Discriminant Analysis*. In general, in the two-group case we fit a linear equation of the type:

Group = a + b1*x1 + b2*x2 +. + bm*xm

where *a* is a constant and *b1* through *bm*are regression coefficients. The interpretation of the results of a two-group problem is straightforward and closely follows the logic of multiple regression: Those variables with the largest (standardized) regression coefficients are the ones that contribute most to the prediction of group membership.

Interpreting the Discriminant Functions

As before, we will get *b* (and standardized *beta*) coefficients for each variable in each discriminant (now also called *canonical*) function, and they can be interpreted as usual: the larger the standardized coefficient, the greater is the contribution of the respective variable to the discrimination between groups. However, these coefficients do not tell us between which of the groups the respective functions discriminate. We can identify the nature of the discrimination for each discriminant (canonical) function by looking at the means for the functions across groups. We can also visualize how the two functions discriminate between groups by plotting the individual scores for the two discriminant functions

Significance of Discriminant Functions

One can test the number of roots that add *significantly* to the discrimination between group. Only those found to be statistically significant should be used for interpretation; non-significant functions (roots) should be ignored.

Classification (Predictive Discriminant Analysis)

Another major purpose to which discriminant analysis is applied is the issue of predictive classification of cases. Once a model has been finalized and the discriminant functions have been derived, how well can we *predict* to which group a particular case belongs?

A Priori and post hoc Predictions

Before going into the details of different estimation procedures, we would like to make sure that this difference is clear. Obviously, if we estimate, based on some data set, the discriminant functions that best discriminate between groups, and then use the *same* data to evaluate how accurate our prediction is, then we are very much capitalizing on chance. In general, one will *always* get a worse classification when predicting cases that were not used for the estimation of the discriminant function. Put another way, *post hoc* predictions are always better than *a priori* predictions. Therefore, one should never base one's confidence regarding the correct classification of future observations on the same data set from which the discriminant functions were derived; rather, if one wants to classify cases predicatively, it is necessary to collect new data to "try out" (cross-validate) the utility of the discriminant functions.

Classification Functions

These are not to be confused with the discriminant functions. The classification functions can be used to determine to which group each case most likely belongs. There are as many classification functions as there are groups. Each function allows us to compute *classification scores* for each case for each group, by applying the formula:

$$Si = ci + wi1*x1 + wi2*x2 +. + wim*xm$$

In this formula, the subscript *i* denotes the respective group; the subscripts *1, 2,., m* denote the *m* variables; *ci* is a constant for the *i*'th group, *wij*is the weight for the *j*'th variable in the computation of the classification score for the *i*'th group; *xj*is the observed value for the respective case for the *j*'th variable. *Si* is the resultant classification score. We can use the classification functions to directly compute classification scores for some new observations.

Classification of Cases

Once we have computed the classification scores for a case, it is easy to decide how to classify the case: in general we classify the case as belonging to the group for which it has the highest classification score. Thus, if we were to study high school students' post-school career/educational choices (*e.g.*, attending college, attending a professional or trade school, or getting a job) based on several variables assessed one year prior to graduation, we could use the classification functions to predict what each student is most likely to do after SSLC. However, we would also like to know the *probability* that the student will make the predicted choice. Those probabilities are called *posterior* probabilities, and can also be computed. However, to understand how those probabilities are derived, let us first consider the so-called *Mahalanobis* distances.

Mahalanobis Distances

In general, the Mahalanobis distance is a measure of distance between two points in the space defined by two or more correlated variables. For example, if there are two variables that are uncorrelated, then we could plot points (cases) in a standard two-dimensional scatter plot; the Mahalanobis distances between the points would then be identical to the Euclidean distance; that is, the distance as, for example, measured by a ruler. If there are three uncorrelated variables, we could also simply use a ruler (in a 3-D plot) to determine the distances between points. If there are more than 3 variables, we cannot represent the distances in a plot any more. Also, when the variables are correlated, then the axes in the plots can be thought of as being *non-orthogonal*; that is, they would not be positioned in right angles to each other. In those cases, the simple Euclidean distance is not an appropriate measure, while the Mahalanobis distance will adequately account for the correlations.

Mahalanobis Distances and Classification

For each group in our sample, we can determine the location of the point that represents the means for all variables in the multivariate space defined by the variables in the model. These points are called group *centroids*. For each case we can then compute the Mahalanobis distances (of the respective case) from each of the group centroids. Again, we would classify the case as belonging to the group to which it is closest, that is, where the Mahalanobis distance is smallest.

Posterior Classification Probabilities

Using the Mahalanobis distances to do the classification, we can now derive probabilities. The probability that a case belongs to a particular group is basically proportional to the Mahalanobis distance from that group centroid. In summary, the posterior probability is the probability, based on our knowledge of the values of other variables that the respective case belongs to a particular group.

Summary of the Prediction

A common result that one looks at in order to determine how well the current classification functions predict group membership of cases is the *classification matrix*. The classification matrix shows the number of cases that were correctly classified (on the diagonal of the matrix) and those that were misclassified.

4.2.8 Cluster Analysis

The purpose of cluster analysis is to reduce a large data set to meaningful subgroups of individuals or objects. The division is accomplished on the basis of similarity of the objects across a set of specified characteristics. Outliers are a problem with this technique, often caused by too many irrelevant variables. The sample should be representative of the population, and it is desirable to have uncorrelated factors. There are three main clustering methods: hierarchical, which is a treelike process appropriate for smaller data sets; nonhierarchical, which requires specification of the number of clusters a priori; and a combination of both. There are four main rules for developing clusters: the clusters should be different, they should be reachable, they should be measurable, and the clusters should be profitable (big enough to matter). This is a great tool for market segmentation.

4.2.9 Multidimensional Scaling (MDS)

The purpose of MDS is to transform consumer judgments of similarity into distances represented in multidimensional space. This is a decompositional approach that uses perceptual mapping to present the dimensions. As an exploratory technique, it is useful in examining unrecognized dimensions about products and in uncovering comparative evaluations of products when the basis for comparison is unknown. Typically there must be at least four times as many objects being evaluated as dimensions. It is possible to evaluate the objects with non-metric preference rankings or metric similarities (paired comparison) ratings. Kruskal's Stress measure is a "badness of fit" measure; a stress percentage of 0 indicates a perfect fit, and over 20 per cent is a poor fit.

The dimensions can be interpreted either subjectively by letting the respondents identify the dimensions or objectively by the researcher.

4.2.10 Correspondence Analysis

This technique provides for dimensional reduction of object ratings on a set of attributes, resulting in a perceptual map of the ratings. However, unlike MDS, both independent variables and dependent variables are examined at the same time. This technique is more similar in nature to factor analysis. It is a compositional technique, and is useful when there are many attributes and many companies. It is most often used in assessing the effectiveness of advertising campaigns. It is also used when the attributes are too similar for factor analysis to be meaningful. The main structural approach is the development of a contingency (crosstab) table. This means that the form of the variables should be nonmetric. The model can be assessed by examining the Chi-square value for the model. Correspondence analysis is difficult to interpret, as the dimensions are a combination of independent and dependent variables.

4.2.11 Conjoint Analysis

Conjoint analysis is often referred to as "trade-off analysis," since it allows for the evaluation of objects and the various levels of the attributes to be examined. It is both a compositional technique and a dependence technique, in that a level of preference for a combination of attributes and levels is developed. A part-worth, or utility, is calculated for each level of each attribute, and combinations of attributes at specific levels are summed to develop the overall preference for the attribute at each level. Models can be built that identify the ideal levels and combinations of attributes for products and services.

4.2.12 Time Series Analysis

Time series has a long history in social sciences, especially in economics and finance. As it is well known, much of economics and finances are concerned with modeling dynamics and systematization of data over time was a subject that appeared early. The cumulated historical data permitted to applied statistical methods in order to find evidence of causation between social variables, finding some support to social theories. Considering the non-experimental nature of the social sciences, this also encourages the development of statistical techniques. For this reason, much of the statistical effort, in particular econometric effort was focused on developing powerful statistical tests, considering the availability of small samples. The basic time series modeling, covering univariate time series analysis utilizing the Box-Jenkins approach, along with time series analysis within a classical regression framework, multivariate time series analysis, focusing first on vector autoregression had gained momentum over a period of time. Recently, several advanced techniques like ARIMA models, time-series regression, unit-root diagnosis, vector autoregressive models, error-correction models, intervention models, fractional integration, ARCH models, structural breaks, and forecasting has been playing a lead role in determining the dynamics of any time series data.

ARIMA (p,d,q) Forecasting Equation

ARIMA models are, in theory, the most general class of models for forecasting a time series which can be made to be "stationary" by differencing (if necessary), perhaps in conjunction with non-linear transformations such as logging or deflating (if necessary). The acronym ARIMA stands for Auto-Regressive Integrated Moving Average. Lags of the stationarized series in the forecasting equation are called "autoregressive" terms, lags of the forecast errors are called "moving average" terms, and a time series which needs to be differenced to be made stationary is said to be an "integrated" version of a stationary series. Random-walk and random-trend models, autoregressive models, and exponential smoothing models are all special cases of ARIMA models. A nonseasonal ARIMA model is classified as an "ARIMA (p,d,q)" model, where:

- **p** is the number of autoregressive terms,
- **d** is the number of non-seasonal differences needed for stationarity, and
- **q** is the number of lagged forecast errors in the prediction equation.

In terms of y, the general forecasting equation is:

$$\hat{y}_t = \mu + \phi_1 y_{t-1} + \ldots + \phi_p y_{t-p} - \theta_1 e_{t-1} - \ldots - \theta_q e_{t-q}$$

Here the moving average parameters (θ's) are defined so that their signs are negative in the equation, following the convention introduced by Box and Jenkins. Some authors and software (including the R programming language) define them so that they have plus signs instead. When actual numbers are plugged into the equation, there is no ambiguity, but it's important to know which convention your software uses when you are reading the output. Often the parameters are denoted there by AR(1), AR(2), ..., and MA(1), MA(2), ... *etc.*To identify the appropriate ARIMA model for Y, you begin by determining the order of differencing (d) needing to stationarize the series and remove the gross features of seasonality, perhaps in conjunction with a variance-stabilizing transformation such as logging or deflating. If you stop at this point and predict that the differenced series is constant, you have merely fitted a random walk or random trend model. However, the stationarized series may still have autocorrelated errors, suggesting that some number of AR terms ($p \geq 1$) and/or some number MA terms ($q \geq 1$) are also needed in the forecasting equation.

4.2.13 Artificial Neural Networks

ANNs are computing systems inspired by the biological neural network that constitute animal brains. Such systems learn (progressively improve performance on) tasks by considering examples, generally without task-specific programming. An ANN is based on a collection of connected units or nodes called artificial neurons (analogous to biological neurons in an animal brain). Each connection (analogous to a synapse) between artificial neurons can transmit a signal from one to another. The artificial neuron that receives the signal can process it and then signal artificial neurons connected to it. In common ANN implementation, the signal at a connection between artificial neurons is a real number, and the output of each artificial neuron is calculated by a non-linear function of the sum of its inputs. Artificial neurons and connections typically have a weight that adjusts as learning proceeds. The weight increases or decreases the strength of the signal at a connection. Artificial neurons may have a threshold such that only if the aggregate signal crosses that threshold is the signal sent. Typically, artificial neurons are organized in layers. Different layers may perform different kinds of transformations on their inputs. Signals travel from the first (input), to the last (output) layer, possibly after traversing the layers multiple times. The original goal of the ANN approach was to solve problems in the same way that a human brain would. Over time, attention focused on matching specific mental abilities, leading to deviations from biology. ANNs have been used on a variety of tasks, including computer vision, speech recognition, machine translation, social network filtering, playing board and video games and medical diagnosis.

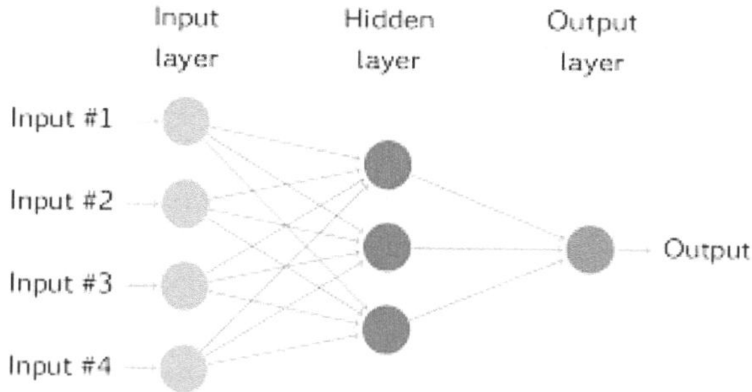

4.2.14 Structural Equation Modelling (SEM)

Unlike the other multivariate techniques discussed, structural equation modelling (SEM) examines multiple relationships between sets of variables simultaneously. This represents a family of techniques, including LISREL, latent variable analysis, and confirmatory factor analysis. SEM can incorporate latent variables, which either are not or cannot be measured directly into the analysis. For example, intelligence levels can only be inferred, with direct measurement of variables like test scores, level of education, grade point average, and other related measures. These tools are often used to evaluate many scaled attributes or to build summated scales.

4.2.15 Artificial Intelligence (AI)

Artificial intelligence is imparting a cognitive ability to a machine. The benchmark for AI is the human intelligence regarding reasoning, speech, and vision. This benchmark is far off in the future. AI has three different levels:

1. **Narrow AI**: An artificial intelligence is said to be narrow when the machine can perform a specific task better than a human. The current research of AI is here now
2. **General AI**: An artificial intelligence reaches the general state when it can perform any intellectual task with the same accuracy level as a human would
3. **Active AI**: An AI is active when it can beat humans in many tasks

Early AI systems used pattern matching and expert systems. Artificial intelligence is imparting a cognitive ability to a machine. Early AI systems used pattern matching and expert systems. The idea behind machine learning is that the machine can learn without human intervention. The machine needs to find a way to learn how to solve a task given the data. Deep learning is the breakthrough in the field of artificial intelligence. When there is enough data to train on, deep learning achieves impressive results, especially for image recognition and text translation. The main reason is the feature extraction is done automatically in the different layers of the network.

4.2.15.1 What is Machine Learning (ML)?

As the name suggests, machine learning can be loosely interpreted to mean empowering computer systems with the ability to "learn". The intention of ML is to enable machines to learn by themselves using the provided data and make accurate predictions.ML is a subset of artificial intelligence; in fact, it's simply a technique for realizing AI. It is a method of training algorithms such that they can learn how to make decisions raining in machine learning entails giving a lot of data to the algorithm and allowing it to learn more about the processed information. For example, here is a table that identifies the type of fruit based on its characteristics:

Weight (grams)	Texture	Type of Fruit
155	Rough	Orange
180	Rough	Orange
135	Smooth	Apple
110	Smooth	Apple
120	Smooth	?

As you can see on the table above, the fruits are differentiated based on their weight and texture. However, the last row gives only the weight and texture, without the type of fruit. And, a machine learning algorithm can be developed to try to identify whether the fruit is an orange or an apple. After the algorithm is fed with the training data, it will learn the differing characteristics between an orange and an apple. Therefore, if provided with data of weight and texture, it can predict accurately the type of fruit with those characteristics.

4.2.15.2 What is Deep Learning (DL)?

As earlier mentioned, deep learning is a subset of ML; in fact, it's simply a technique for realizing machine learning. In other words, DL is the next evolution of machine learning. DL algorithms are roughly inspired by the information processing patterns found in the human brain. Just like we use our brains to identify patterns and classify various types of information, deep learning algorithms can be taught to accomplish the same tasks for machines. The brain usually tries to decipher the information it receives. It achieves this through labeling and assigning the items into various categories. Whenever we receive new information, the brain tries to compare it to a known item before making sense of it — which are the same concept deep learning algorithms employ. For example, artificial neural networks (ANNs) are a type of algorithms that aim to imitate the way our brains make decisions. Comparing deep learning vs machine learning can assist you to understand their subtle differences. For example, while DL can automatically discover the features to be used for classification, ML requires these features to be provided manually. Furthermore, in contrast to ML, DL needs high-end machines and considerably big amounts of training data to deliver accurate results.

4.2.15.3 Areas of Application of AI in Agriculture

There are huge applications of AI in agriculture starting from field preparation to harvesting of crops. There are AI based technologies/innovations that are able to correctly assess the field layout, and guide a tractor (with or without a driver) to prepare the field in such a manner so as to maximize production. These technologies can work without a driver, and accurately assess the lay of the land to maximize land usage. AI-based robotic applications have now made it possible to sow seeds in a straight line with adequate space between them, replacing seed-drills and manual sowing. Pesticides, herbicides, and other agrochemicals are usually sprayed all over the plants and sometimes even randomly around un-affected areas. Current innovations in image-based analytics have made it possible for drone-based sprayers to target infected areas of the plant.

New age AI-based apps have made it possible for farmers to click photos of crops that are infected with diseases and get a probability score of the infected disease on a real-time basis.As in the case of diagnosing diseases, here too, the farmer has to click photographs of crops to gets a probability score of nutrient deficiency on a real-time basis. However, a physical test is recommended for correct diagnosis. Now it is possible for users to map the lay of the land basis the existing soil nutrients. This is very useful for the application of Urea, DAP,

MOP, SSP, and separate doses of N, P and K which can be appropriately applied on the ground without compromising soil health. Technologies to monitor water release timing, flow, quantum, and speed ensure that water is used most productively basis past data of soil health and soil nutrients. There are technologies that look at environmental factors and soil conditions to predict yields for the coming seasons. There are also image-based technologies that look at images of plantations – palm, rubber, sugarcane, tea, coffee, coconuts, apples, mangoes, *etc.* Machines can now harvest crops more precisely and more cleanly, thanks to AI. This is again done by looking at images and cutting crops so as to maximize yield. AI-based technologies now make it possible for farmers and companies to take photographs of commodities with proper displaying of details on the screen. These results show the count and the weight of the commodity with a fair degree of accuracy. Combinations of above technologies on weather, yield, and quality can be used by insurance company as a reliable data-driven measure to tailor customized insurance policies. Correct processing mixes for both food and feed items can be obtained effortlessly with AI. These mixes may be further customized, basis the animal type, stage of growth, diet preferences, *etc.* Results of the above technologies especially on yield and quality can be shared by banks to take loans. Banks can, in turn, use these data as a basis for funding, given the reliability and retrieval-ability. AI and IoT for Agricultural Applications are given below:

Precision Farming with the help of Micro Sensors

Precision farming require a keen eye, as farmers scan/monitor their crops for symptoms of disease and contamination. The technology is an application under machine learning development services can display stats remotely via a smart phone or similar device, allowing farmers to see instant alerts about what's happening in their fields, whether it has to do with pests, disease, or something else.

Drone Operations

Crop Monitoring by using cameras and sensors embedded inside, the drones gather farm information about plants down to a single leaf.

IoT based Sensors

By combining AI control solutions and IoT sensors, farmers can better protect crops from pests. At the same time, fewer chemicals enter the surrounding environment, including the soil underneath.

Predictive Analytics

With the help of AI, farm data like temperature, weather conditions, soil conditions, and soil usage can be effectively collected and analyzed to influence decision making on the farm which makes best possible crop suggestion before the season keeping the past data and other sensor, environment data in the view.

Over the years, the nature has experienced a drastic increase in the pollution level and unpredictable climatic conditions. The change in climate has made it difficult for farmers to determine the right time for sowing seeds and that's where AI comes into the picture. With the help of AI it is possible to monitor how weather, seasonal sunlight, wind speed, and rain will affect the crop planting cycles. Weather forecasting will help farmers to know when the seeds should be sown. IBM weather forecast alerts in case of any weather hazards and provides possible solutions to minimize environmental impact and reduce costs.

Digital Monitoring of Soil and Crop

Soil determines the growth and health of any crop. The nutrition of the soil is directly related with the quality of the crop. Deforestation deteriorates the soil quality over time and it becomes a challenge for the farmers to understand what soil will be best for what crop. An AI-based device *viz.*, Plantix uses image recognition that helps farmers to identify nutrient deficiencies in soil, plant pests and other diseases. With the help of Plantix, farmers can easily figure out which fertilizer to use in order to improve the quality of the harvest. The app also provides tips and solutions for the detected problems.

Using Drones for Data Collection

The crop yield can be improved by the application of AI and machine learning in agriculture through real-time sensor data and visual analytics data from drones. Drones can make video surveillance and can analyze growth patterns of the crops. On the other hand, sensors can store data of moisture, fertilizer and natural nutrient levels. For which, drones are considered as reliable tool for capturing data on how specific fertilizers, watering patterns and pesticide treatment methods are improving crop yields. Precision Hawk allows farmers to virtually walk around their fields with the help of drones. Farming operations of all sizes either big or small, are using drones to reduce the time and costs associated with crop data collection.

Precision Farming

Use of AI in agriculture provides farmers with proper guidance regarding soil and water management, suitable crop rotation, precise harvesting, optimum planting, minimal pest attacks, *etc.* Machine learning algorithms with the images captured by satellites and drones, can predict weather conditions, nutrition management and analyze crop sustainability. Farmers with smart phones and AI applications can get a customized plan for their lands. The IoT based and AI driven solutions can increase food sustainably, grow production and revenues without depleting any of our natural resources.

AI Powered Robots

AI based robots able to perform multiple farming tasks. These robots can be trained to control the excessive growth of weeds and to harvest crops. They will also be trained to collect and pack the harvested crops. The goal is to replace the manual tasks and accomplish them at a much faster rate with larger volumes when compared to humans. Agrobot robots have the ability to operate 24/7 which increases the efficiency, optimizes the cost of precision to weed, hoe, and harvest. The Argobot E-series with advanced AI system not only picks up the crops but it can also identify the ripeness of the crop in the field.

AI Enabled Surveillance Systems

AI based technology can use the satellite images in order to compare them with past data to detect animal or human breaches. This minimizes the risk of any domestic or wild animal that can destroy crops. AI algorithms can identify something as small as an insect, for example- locust, grasshopper, *etc.* Once an intrusion is detected, alerts are immediately sent to the farmers on their smart phones so that they can take the required precautions. Plantix app turns your smart phone into a mobile crop doctor with which you can accurately detect pests and diseases on crops within seconds. Plantix serves as a complete solution for crop production and management.

REFERENCES

Betsy J. Becker (499), College of Education, Michigan State University, Anderson. T W (1958) An Introduction to Multivariate Statistical Analysis. John Wiley and Sons, London.

Borg I. and Groenen P. (1997).Modern multidimensional scaling. New York: Springer-Verlag.] Escoer, B., and Pagµes, J. (1988).Analyses factorielles multiples. Paris: Dunod.

EBetsy J. Becker (499), College of Education, Michigan State University, East Lansing, Michigan 48824.

Fisher. R A (1936) The use of multiple measurements in taxonomic problems. Annals of Eugenics 7.179-188.

Flury,Bernhard and Riedwyl, Hans (1988) Multivariate Statistics:A Practical *Approach.*Chapman and Hall, London.

Gnanadesikan, R (1977), Methods for Statistical Data Analysis of Multivariate Observations.

Gould, S J (1981) The Misrneasure of Man.W WNorton and Co, New York.

Johnson R.A. and Wichern D.W. (2002).Applied multivariate statistical anal-ysis. Upper Saddle River (NJ): Prentice-Hall.

Kaciak, Eugene and Koczkodaj, Waldemar W (1989) A spreadsheet approach to principal components analysis. Journal of Microcomputer Applications 12, 281-291.

LisabethF. DiLalla (439), School of Medicine, Southern Illinois University, Carbondale, Illinois 62901

Mark L. Davison (323), Department of Educational Psychology, University of Minnesota, Minneapolis, Minnesota 55455

Michael T. Brown (209), Graduate School of Education, University of California, Santa Barbara, Santa Barbara, California 93016

Naes T., Risvik E. (Eds.) (1996). Multivariate analysis of data in sensory science. New York: Elsevier.

Rao, C R (1964) The use and interpretation of principal component analysis in applied research. ***Sankya A*** 26, 329-358.

ReneV. Dawis (65), University of Minnesota, Minneapolis, Minnesota 55414; and The Ball Foundation, Glen Ellyn, Illinois 60137

Robert Cudeek (265), Department of Psychology, University of Minnesota, Minneapolis, Minnesota 55455

Roessner, D. (2002). *Outcome Measurement in the United States: State of the Art.* Paper presented at the Annual Meeting of the American Association for the Advancement of Science, Boston, MA.

Scheirer, M.A. (2000). Getting more "bang" for your performance measures "buck". *American Journal of Evaluation, 21,* 139-149.

Steven D. Brown (3), Department of Leadership, Foundations, and Counseling Psychology, Loyola University of Chicago, Wilmettc, Illinois 60091

Weller S.C. and Romney A.K. (1990). Metric scaling: Correspondence analysis. Thou.

— Chapter 5 —

Non-linear Programming for Research Data

A linear programming problem aims to maximize or minimize a linear function subject to the linear constraints. But in many cases, the objective function may not be linear or all/some of the constraints may not be linear or both the objective and the constraints may not be linear. Such an optimization problem is called a non-linear programming problem (NLPP). In this chapter, we discuss about NLPP's. Non-linear programming is the process of solving an optimization problem defined by a system of equalities and inequalities, collectively termed constraints, over a set of unknown real variables, along with an objective function to be maximized or minimized, where some of the constraints or the objective function are non-linear. It is the sub-field of mathematical optimization that deals with problems that are not linear. A typical non-convex problem is that of optimizing transportation costs by selection from a set of transportation methods, one or more of which exhibit economies of scale, with various connectivity and capacity constraints. An example would be petroleum product transport given a selection or combination of pipeline, rail tanker, road tanker, river barge. Owing to economic batch size the cost functions may have discontinuities in addition to smooth changes. In experimental science, some simple data analysis (such as fitting a spectrum with a sum of peaks of known location and shape but unknown magnitude) can be done with linear methods, but in general these problems, also, are non-linear. Typically, one has a theoretical model of the system under study with variable parameters in it and a model the experiment or experiments, which may also have unknown parameters. One tries to find a best fit numerically. In this case one often wants a measure of the precision of the result, as well as the best fit itself. Non-linear programming (NP) involves minimizing or maximizing a non-linear objective function subject to bound constraints, linear constraints, or non-linear constraints, where the constraints can be inequalities or equalities. Example problems in engineering include analyzing design tradeoffs, selecting optimal designs, and computing optimal trajectories. Unconstrained non-linear programming is the mathematical problem of finding a vector x that is a local minimum to the non-linear scalar function f(x).

5.1 Basics of Non-linear Programming

5.1.1 Definition

Let n, m, and p be positive integers. Let X be a subset of R^n, let f, g_i, and h_j be real-valued functions on X for each i in $\{1, \ldots, m\}$ and each j in $\{1, \ldots, p\}$, with at least one of f, g_i, and h_j being non-linear. A non-linear minimization problem is an optimization problem of the form

{\displaystyle {\begin{aligned}{\text{minimize }} and f(x)\\{\text{subject to }} and g_{i}(x)\leq 0{\text{ for each }}i\in \{1,\dotsc,m\}\\ and h_{j}(x)=0{\text{ for each }}j\in \{1,\dotsc,p\}\\ and x\in X.\end{aligned}}}Minimize f(x)

Subject to $g_i(x) \le$ for each i€{1,…,m}

$h_j(x)=0$ for each j€{1,…,p}

x€X

A non-linear maximization problem is defined in a similar way.

5.1.2 Different Types of Constraint Sets

- There are several possibilities for the nature of the constraint set, also known as the feasible set or feasible region.
- An *infeasible* problem is one for which no set of values for the choice variable satisfies all the constraints. That is, the constraints are mutually contradictory, and no solution exists; the feasible set is the empty set.
- A *feasible* problem is one for which there exists at least one set of values for the choice variables satisfying all the constraints.
- An *unbounded* problem is a feasible problem for which the objective function can be made to be better than any given finite value. Thus there is no optimal solution, because there is always a feasible solution that gives a better objective function value than does any give proposed solution.

5.1.3 Methods

- If the objective function f is linear and the constrained space is a polytope, the problem is a linear programming problem, which may be solved using well-known linear programming techniques such as the simplex method.
- If the objective function is concave (maximization problem), or convex (minimization problem) and the constraint set is convex, then the program is called convex and general methods from convex optimization can be used in most cases.
- If the objective function is quadratic and the constraints are linear, quadratic programming techniques are used.
- If the objective function is a ratio of a concave and a convex function (in the maximization case) and the constraints are convex, then the problem can be transformed to a convex optimization problem using fractional programming techniques.

Several methods are available for solving non-convex problems. One approach is to use special formulations of linear programming problems. Another method involves the use of branch and bound techniques, where the program is divided into subclasses to be solved with convex (minimization problem) or linear approximations that form a lower bound on the overall cost within the subdivision. With subsequent divisions, at some point an actual solution will be obtained whose cost is equal to the best lower bound obtained for any of the approximate solutions. This solution is optimal, although possibly not unique. The algorithm may also be stopped early, with the assurance that the best possible solution is within a tolerance from the best point found; such points are called ε-optimal. Terminating to ε-optimal points is typically necessary to ensure finite termination. This is especially useful for large, difficult problems and problems with uncertain costs or values where the uncertainty can be estimated with appropriate reliability estimation. Under differentiability and constraint qualifications, the Karush–Kuhn–Tucker (KKT) conditions provide necessary conditions for a solution to be optimal. Under convexity, these conditions are also sufficient. If some of the functions are non-differentiable, sub differential versions of Karush–Kuhn–Tucker (KKT) conditions are available.

Local Optima and Global Optima

A local minimum of a function is a point where the function value is smaller than at nearby points, but possibly greater than at a distant point. A global minimum is a point where the function value is smaller than at all other feasible points.

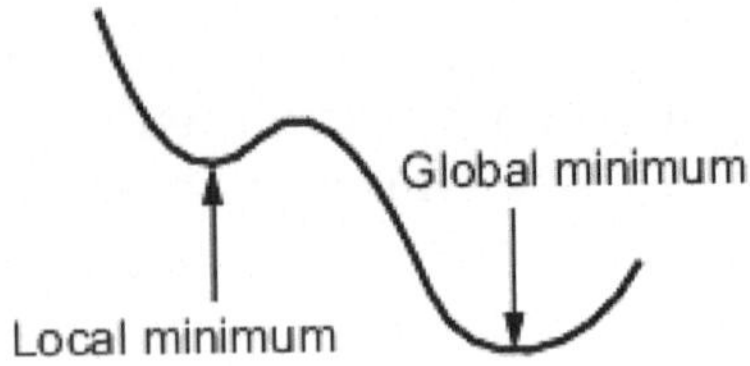

If you need a global optimum, you must find an initial value for your solver in the basin of attraction of a global optimum. Suggestions for ways to set initial values to search for a global optimum:

- Use a regular grid of initial points.
- Use random points drawn from a uniform distribution if your problem has all its coordinates bounded. Use points drawn from normal, exponential, or other random distributions if some components are unbounded. The less you know about the location of the global optimum, the more spread-out your random distribution should be. For example, normal distributions rarely sample more than three standard deviations away from their means, but a Cauchy distribution (density $1/(\pi(1 + x^2)$) makes hugely disparate samples.
- Use identical initial points with added random perturbations on each coordinate, bounded, normal, exponential, or other.

2-dimensional Example

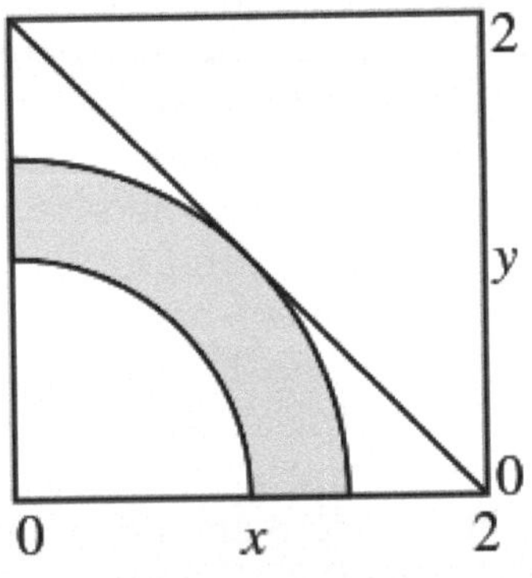

The tangency of the line with the constrained space represents the solution. The line is the best achievable contour line (locus with a given value of the objective function).A simple problem can be defined by the constraints

$x_1 \geq 0$

$x_2 \geq 0$

$x_1^2 + x_2^2 \geq 1$

$x_1^2 + x_2^2 \leq 2$

with an objective function to be maximized

$f(\mathbf{x}) = x_1 + x_2$

where $\mathbf{x} = (x_1, x_2)$.

3-dimensional Example

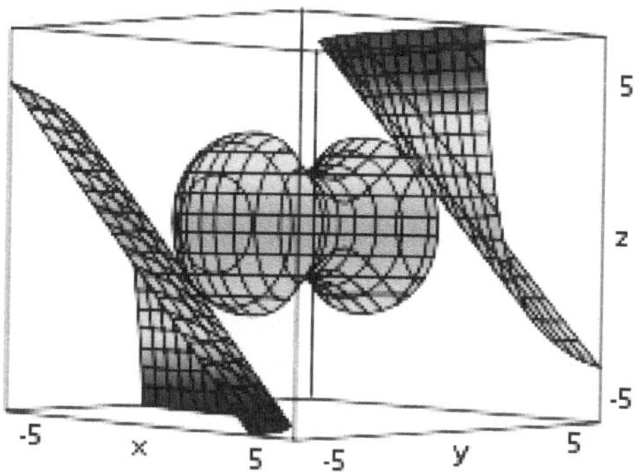

The tangency of the top surface with the constrained space in the centre represents the solution. Another simple problem can be defined by the constraints

$x_1^2 - x_2^2 + x_3^2 \leq 2$

$x_1^2 + x_2^2 + x_3^2 \leq 10$

with an objective function to be maximized

$f(\mathbf{x}) = x_1 x_2 + x_2 x_3$

where $\mathbf{x} = (x_1, x_2, x_3)$.

Given a transformation between input and output values, described by a mathematical function *f*, optimization deals with generating and selecting a best solution from some set of available alternatives, by systematically choosing input values from within an allowed set, computing the output of the function, and recording the best output values found during the process. Many real-world problems can be modeled in this way. For example, the inputs can be design parameters of a motor, the output can be the power consumption, or the inputs can be business choices and the output can be the obtained profit. The use of optimization software requires that the function *f* is defined in a suitable programming language and connected at compile or run time to the optimization software. The optimization software will deliver input values in *A*, the software module realizing *f* will deliver the computed value *f*(*x*) and, in some cases, additional information about the function like derivatives. Some of the common used software are AIMMS. Analytic Solver, ASTOS, CPLEX, General Algebraic Modeling System, LINDO, MATLAB *etc.*

5.2 Application of Non-linear Programming

5.2.1 Application of Derivatives to Non-linear Programming for Prescriptive Analytics

Besides the traditional probability theory and statistics, current machine learning techniques work in complete sync with linear algebra, graph theory, dynamic programming, multivariate calculus *etc.* As far as multivariate calculus is concerned, the different methods that lend support to machine learning algorithms are differential and integral calculus, partial derivatives, gradient and directional derivative, vector-valued function, Jacobian matrix and determinant, Hessian matrix, Laplacian and Lagrangian distributions *etc.* This blog will discuss the applications of second order derivatives and partial derivatives on optimization problems, as required for prescriptive analytics. Prescriptive analytics provides precise decisions on the course of action necessary for future success. One of the prominent applications of prescriptive analytics in marketing is the optimization problem of marketing budget allocation. The business problem is to figure out the optimum quantity of budget that needs to be allocated from the total advertising budget to each of the advertising media like TV, press, internet video *etc.* for maximizing the revenue. The budget optimization problem is solved either through Linear or Non-linear Programming (NLP) which depends on whether:

1. The objective function is linear/non-linear
2. The feasible region is determined by linear/non-linear constraints

Thus, one of the important assumptions for linear programming is the constant returns to scale for each of the advertising media. Partial derivative is the other prominent application of calculus on optimization problems. Partial derivatives is a function with several variables which are accomplished when a particular variable's derivative is computed keeping other variables constant. One of the most widely used applications of partial derivative is the least square criterion where the objective is to find out the best fitting line by minimizing the distance of the line from the data points. This is achieved by setting first order partial derivatives of the intercepts and the slopes equal to zero. The other major applications of partial derivatives are as follows:

- ☆ The second order partial derivative is used in an optimization problem to figure out whether a given critical point is a relative maximum, relative minimum, or a saddle point.
- ☆ The assignment of penalty for conversion of an optimization problem to an unconstrained optimization problem through sequential unconstrained minimization technique.

The approaches discussed show how calculus can be integrated with non-linear programming while delivering an optimized solution. In the same manner, Lagrangean-based techniques can also be integrated with Mixed Integer Non-Linear Programming (MINLP) to provide with the marketing budget optimization solution. Currently, data scientists cannot bank on a single technique to provide analytics solution. The real challenge is to figure out how multiple techniques can be creatively combined to provide a solution as unique as the business problem.

5.2.2 Application of Non-linear Programming Models in Agriculture

Experience with traditional programming models shows that a considerable improvement in performance is possible by adequately incorporating non-linear relationships. Particular emphasis will be given to the calibration and validation problems involved in this type of model. With the help of the Turkish agricultural sector model (TASM) it will be demonstrated that an empirical specification of a non-linear programming model for the agricultural sector is possible even with poor statistical data and that an operational model version can be handled.

A non-linear programming model may be used for optimal planning and best use of nutrient ingredients. The solution of non-linear problems using penalization methods, which transforms the problem into an unconstrained non-linear optimization is oriented towards the study of Multi objective Optimization Problems. The production planning for examining the process of allocating land on agriculture activities and determining the optimal cropping patterns to achieve maximum in the income, and minima of irrigation water, use of fertilizers and number of workers non-linear programming may be used.

The discipline of agricultural economics has played a pioneering role in the application of mathematical models in economics. Originally, farm management specialists developed enterprise budgets to help farmers identify their best farming activities. These budgets, which specified the amounts of inputs required and outputs generated by allocating a unit of land to various crop or livestock production activities, facilitated the application of mathematical programming techniques as they were developed to offer normative advice to farmers. Eventually, programming models were expanded to offer advice at regional and national levels on issues as diverse as commercial agricultural policy, environmental policy, water and soil conservation policy, and public investment in infrastructure such as waterway and irrigation development. Agriculture was a natural application of stochastic and dynamic programming because of the stochastic nature of production and prices and the dynamic issues of capital investment and grain stock accumulation. On another front, agricultural economics played a key role in the development and application of econometrics for positive modeling purposes. With a wide variety of reasonably competitive product markets and abundant public data, agriculture presented a ready field of application for various statistical regression techniques as they were developed. Simultaneous equation models were easily expanded to address issues of international trade as they became important. In many cases, agricultural production characteristics such as heterogeneity, imperfect anticipation of product prices, and production risk motivated development and refinement of econometric techniques. With perceived competitive conditions, agriculture served as a ready laboratory for application of techniques that increasingly incorporated microeconomic theory in model specification, first with primal and then with dual approaches. In more recent times, as contracting has replaced open market transactions, agriculture has become a ready laboratory for application of game theory and the principal-agent models of mechanism design that now dominate microeconomic analysis.

Dynamic programming (DP) has shown itself to be an appropriate methodology for devising strategies of fisheries management. The model by Stanlfel (1988) has several goals in the same area: to elucidate the process of formulation and its requirements; to illustrate how practical constraints may be incorporated into the problems; to examine the realities of computation; to show that models of other than an equilibrium nature may be of use and that finite horizon models are more realistic; to emphasize the relevance of some models, not previously employed; and to correct some abuses and misconceptions. He also made some effort to reveal practical benefits to real-world decision-making in this problem area occasioned by the nature of DP. Certain significant advantages have been overlooked by previous researches.

Among the model improvements suggested are: backward formulations for more realistic decisions in both the cases of invertible and non-invertible transformations; the introduction of varying catch capacity and associated costs for changes and for underutilized capacity and the computational implications of multispecies models. For elementary deterministic models possible sources of error are identified and recommendations made for dealing with certain of these. A careful presentation was made of the effort involved in handling one-dimensional problems with random variables of known probability distributions. In the absence of distributions adaptive models are mentioned, along with their attending complications and some empirical evidence promoting optimism in that case.

To describe phytoplankton dynamics in the moving water there is a need of the characteristics of mixing and stirring since these processes determine the density distribution of phytoplankton. Since the characteristics of mixing in fluids depend on the control parameters (*e.g.* Reynolds number, *etc.*) many different situations can be studied, and since physical circumstances are so diverse in oceans and freshwaters, almost all possible situations must be studied. On the other hand, competitors form complex hierarchical networks in real ecosystems; networks may be built up by two elementary structures: transitive hierarchy (*e.g.* if B beats A and C beats B, then C beats A), and cyclic hierarchy (*e.g.* B beats A, C beats B, but A beats C). Thus, the dynamics of transitive and cyclical hierarchy either in open and closed chaotic flows (with and without coherent structures) or in turbulent flows to map the possible effects of interaction of population and hydrodynamics in competitive situations can be studied by non-linear models.

REFERENCES

Avriel, Mordecai (2003). *Non-linear Programming: Analysis and Methods.* Dover Publishing. ISBN 0-486-43227-0.

Bazaraa, Mokhtar S. and Shetty, C. M. (1979). *Non-linear programming.Theory and algorithms.* John Wiley and Sons. ISBN 0-471-78610-1.

Bertsekas, Dimitri P. (1999). *Non-linear Programming: 2nd Edition.* Athena Scientific. ISBN 1-886529-00-0.

Bonnans, J. Frédéric; Gilbert, J. Charles; Lemaréchal, Claude*; Sagastizábal, Claudia A. (2006).* Numerical optimization: Theoretical and practical aspects. *Universitext (Second revised ed. of translation of 1997 French ed.). Berlin: Springer-Verlag. pp. xiv+490.*doi:10.1007/978-3-540-35447-5. ISBN 3-540-35445-X. MR 2265882.

Jan Brinkhuis and Vladimir Tikhomirov, *Optimization: Insights and Applications,* 2005, Princeton University Press

Luenberger, David G.; Ye, Yinyu *(2008). Linear and non-linear programming. International Series in Operations Research and Management Science.* ***116*** *(Third ed.). New York: Springer. pp. xiv+546.* ISBN 978-0-387-74502-2. MR 2423726.

Nocedal, Jorge and Wright, Stephen J. (1999). *Numerical Optimization.* Springer. ISBN 0-387-98793-2.

Ruszczyñski, *Andrzej (2006). Non-linear Optimization. Princeton, NJ:* Princeton University Press. *pp. xii+454.* ISBN 978-0691119151. MR 2199043.

S. Bauer and H. Kasnakoglu, Turkish Agricultural Sector Model, Report presented to Ministry of Agriculture, Forestry and Village Affairs, Ankara, 1988

— Chapter 6 —

Measurement and Scaling Techniques Measurement in Research

In social science research, we have to deal with qualitative variables that may further need to undergo reliable measurements and analytical techniques. So, we have to go for scaling of these variables in such a way it can qualify the empirical measurement with reliability and validity.

In our daily life we are said to measure when we use some yardstick to determine weight, height, or some other feature of a physical object. We also measure when we judge how well we like a song, a painting or the personalities of our friends. We, thus, measure physical objects as well as abstract concepts. Measurement is a relatively complex and demanding task, especially so when it concerns qualitative or abstract phenomena. By measurement we mean the process of assigning numbers to objects or observations, the level of measurement being a function of the rules under which the numbers are assigned.

It is easy to assign numbers in respect of properties of some objects, but it is relatively difficult in respect of others. For instance, measuring such things as social conformity, intelligence, or marital adjustment is much less obvious and requires much closer attention than measuring physical weight, biological age or a person's financial assets. In other words, properties like weight, height, *etc.*, can be measured directly with some standard unit of measurement, but it is not that easy to measure properties like motivation to succeed, ability to stand stress and the like. We can expect high accuracy in measuring the length of pipe with a yard stick, but if the concept is abstract and the measurement tools are not standardized, we are less confident about the accuracy of the results of measurement. Technically speaking, measurement is a process of mapping aspects of a domain onto other aspects of a range according to some rule of correspondence. In measuring, we devise some form of scale in the range (in terms of set theory, range may refer to some set) and then transform or map the properties of objects from the domain (in terms of set theory, domain may refer to some other set) onto this scale. For example, in case we are to find the male to female attendance ratio while conducting a study of persons who attend some show, then we may tabulate those who come to the show according to sex.

In terms of set theory, this process is one of mapping the observed physical properties of those coming to the show (the domain) on to a sex classification (the range). The rule of correspondence is: If the object in the domain appears to be male, assign to "0" and if female assign to "1". Similarly, we can record a person's marital status as 1, 2, 3 or 4, depending on whether the person is single, married, widowed or divorced. We can as well record "Yes or No" answers to a question as "0" and "1" (or as 1 and 2 or perhaps as 59 and 60). In this artificial or nominal way, categorical data (qualitative or descriptive) can be made into numerical data and if we thus code the various categories, we refer to the numbers we record as nominal data. Nominal data are numerical in name only, because they do not share any of the properties of the numbers we deal in ordinary arithmetic. For instance if we record marital status as 1,2,3, or 4 as stated above, = we cannot write 4> 2 or 3<4 and we cannot write 3 – 1 = 4 – 2, 1 + 3 = 4 or 4 ¸ 2 = 2. In those situations when we cannot do anything except set up inequalities, we refer to the data as ordinal data. For instance, if one mineral can scratch another, it receives a higher hardness number and on Mohs' scale the numbers from 1 to 10 are assigned respectively to talc, gypsum, calcite, fluorite, apatite, feldspar, quartz, topaz, sapphire and diamond. With these numbers we can write 5 > 2 or 6 < 9 as apatite is harder than gypsum and feldspar is softer than sapphire, but we cannot write for example 10 – 9 = 5 – 4, because the difference in hardness between diamond and sapphire is actually much greater than that between apatite and fluorite. It would also be meaningless to say that topaz is twice as hard as fluorite simply because their respective hardness numbers on Mohs' scale are 8 and 4. The greater than symbol (*i.e.*, >) in connection with ordinal data may be used to designate "happier than" "preferred to" and so on. When in addition to setting up inequalities we can also form differences, we refer to the data as interval data. Suppose we are given the following temperature readings (in degrees Fahrenheit): 58°, 63°, 70°, 95°, 110°, 126° and 135°. In this case, we can write 100° > 70° or 95° < 135° which simply means that 110° is warmer than 70° and that 95° is cooler than 135°. We can also write for example 95°–70° = 135°–110°, since equal temperature differences are equal in the sense that the same amount of heat is required to raise the temperature of an object from 70° to 95° or from 110° to 135°. On the other hand, it would not mean much if we said that 126° is twice as hot as 63°, even though 126° ÷ 63° = 2. To show the reason, we have only to change to the centigrade scale, where the first temperature becomes 5/9 (126 – 32) = 52°, the second temperature becomes 5/9 (63 – 32) = 17° and the first figure is now more than three times the second. This difficulty arises from the fact that Fahrenheit and Centigrade scales both have artificial origins (zeros) *i.e.*, the number 0 of neither scale is indicative of the absence of whatever quantity we are trying to measure. When in addition to setting up inequalities and forming differences we can also form quotients (*i.e.*, when we can perform all the customary operations of mathematics), we refer to such data as ratio data. In this sense, ratio data includes all the usual measurement (or determinations) of length, height, money amounts, weight, volume, area, pressures *etc.* The above stated distinction between nominal, ordinal, interval and ratio data is important for the nature of a set of data may suggest the use of particular statistical techniques*. A researcher has to be quite alert about this aspect while measuring properties of objects or of abstract concepts.

Measurement Scales

From what has been stated above, we can write that scales of measurement can be considered in terms of their mathematical properties. The most widely used classification of measurement scales are:

- ☆ Nominal scale
- ☆ Ordinal scale
- ☆ Interval scale
- ☆ Ratio scale

Nominal Scale

Nominal scale is simply a system of assigning number symbols to events in order to label them. The usual example of this is the assignment of numbers of basketball players in order to identify them. Such numbers cannot be considered to be associated with an ordered scale for their order is of no consequence; the numbers are just convenient labels for the particular class of events and as such have no quantitative value. Nominal scales provide convenient ways of keeping track of people, objects and events. One cannot do much with the numbers involved. For example, one cannot usefully average the numbers on the back of a group of football players and come up with a meaningful value. Neither can one usefully compare the numbers assigned to one group with the numbers assigned to another. The counting of members in each group is the only possible arithmetic operation when a nominal scale is employed. Accordingly, we are restricted to use mode as the measure of central tendency. There is no generally used measure of dispersion for nominal scales. Chi-square test is the most common test of statistical significance that can be utilized, and for the measures of correlation, the contingency coefficient can be worked out.

Nominal scale is the least powerful level of measurement. It indicates no order or distance relationship and has no arithmetic origin. A nominal scale simply describes differences between things by assigning them to categories. Nominal data are, thus, counted data. The scale wastes any information that we may have about varying degrees of attitude, skills, understandings, *etc.* In spite of all this, nominal scales are still very useful and are widely used in surveys and other ex-post-facto research when data are being classified by major sub-groups of the population.

Ordinal Scale

The lowest level of the ordered scale that is commonly used is the ordinal scale. The ordinal scale places events in order, but there is no attempt to make the intervals of the scale equal in terms of some rule. Rank orders represent ordinal scales and are frequently used in research relating to qualitative phenomena. A student's rank in his graduation class involves the use of an ordinal scale. One has to be very careful in making statement about scores based on ordinal scales. For instance, if Ram's position in his class is 10 and Mohan's position is 40, it cannot be said that Ram's position is four times as good as that of Mohan. The statement would make no sense at all. Ordinal scales only permit the ranking of items from highest to lowest. Ordinal measures have no absolute values, and the real differences between adjacent ranks may not be equal. All that can be said is that one person is higher or lower on the scale than another, but more precise comparisons cannot be made. Thus, the use of an ordinal scale implies a statement of 'greater than' or 'less than' (an equality statement is also acceptable) without our being able to state how much greater or less. The real difference between ranks 1 and 2 may be more or less than the difference between ranks 5 and 6. Since the numbers of this scale have only a rank meaning, the appropriate measure of central tendency is the median. A percentile or quartile measure is used for measuring dispersion. Correlations are restricted to various rank order methods. Measures of statistical significance are restricted to the non-parametric methods.

Interval Scale

In the case of interval scale, the intervals are adjusted in terms of some rule that has been established as a basis for making the units equal. The units are equal only in so far as one accepts the assumptions on which the rule is based. Interval scales can have an arbitrary zero, but it is not possible to determine for them what may be called an absolute zero or the unique origin. The primary limitation of the interval scale is the lack of a true zero; it does not have the capacity to measure the complete absence of a trait or characteristic. The Fahrenheit scale is an example of an interval scale and shows similarities in what one can and cannot do with it. One can say that an increase in temperature from 30° to 40° involves the same increase in temperature as an increase from 60° to 70°, but one cannot say that the temperature of 60° is twice as warm as the temperature of 30° because both numbers are dependent on the fact that the zero on the scale is set arbitrarily at the temperature of the freezing point of water. The ratio of the two temperatures, 30° and 60°, means nothing because zero is an arbitrary point. Interval scales provide more powerful measurement than ordinal scales for interval scale also incorporates the concept of equality of interval. As such more powerful statistical measures can be used with interval scales. Mean is the appropriate measure of central tendency, while standard deviation is the most widely used measure of dispersion. Product moment correlation techniques are appropriate and the generally used tests for statistical significance are the 't' test and 'F' test.

Ratio Scale

Ratio scales have an absolute or true zero of measurement. The term 'absolute zero' is not as precise as it was once believed to be. We can conceive of an absolute zero of length and similarly we can conceive of an absolute zero of time. For example, the zero point on a centimeter scale indicates the complete absence of length or height. But an absolute zero of temperature is theoretically unobtainable and it remains a concept existing only in the scientist's mind. The number of minor traffic-rule violations and the number of incorrect letters in a page of type script represent scores on ratio scales. Both these scales have absolute zeros and as such all minor traffic violations and all typing errors can be assumed to be equal in significance. With ratio scales involved one can make statements like "Jyoti's" typing performance was twice as good as that of "Reetu." The ratio involved does have significance and facilitates a kind of comparison which is not possible in case of an interval scale. Ratio scale represents the actual amounts of variables. Measures of physical dimensions such as weight, height, distance, *etc.* are examples. Generally, all statistical techniques are usable with ratio scales and all manipulations that one can carry out with real numbers can also be carried out with ratio scale values. Multiplication and division can be used with this scale but not with other scales mentioned above. Geometric and harmonic means can be used as measures of central tendency and coefficients of variation may also be calculated. Thus, proceeding from the nominal scale (the least precise type of scale) to ratio scale (the most precise), relevant information is obtained increasingly. If the nature of the variables permits, the researcher should use the scale that provides the most precise description. Researchers in physical sciences have the advantage to describe variables in ratio scale form but the behavioral sciences are generally limited to describe variables in interval scale form, a less precise type of measurement.

Sources of Error in Measurement

Measurement should be precise and unambiguous in an ideal research study. This objective, however, is often not met with in entirety. As such the researcher must be aware about the sources of error in measurement. The following are the possible sources of error in measurement.

- ☆ **Respondent:** At times the respondent may be reluctant to express strong negative feelings or it is just possible that he may have very little knowledge but may not admit his ignorance. All this reluctance is likely to result in an interview of 'guesses.' Transient factors like fatigue, boredom, anxiety, *etc.* may limit the ability of the respondent to respond accurately and fully.
- ☆ **Situation:** Situational factors may also come in the way of correct measurement. Any condition which places a strain on interview can have serious effects on the interviewer-respondent rapport. For instance, if someone else is present, he can distort responses by joining in or merely by being present. If the respondent feels that anonymity is not assured, he may be reluctant to express certain feelings.
- ☆ **Measurer:** The interviewer can distort responses by rewording or reordering questions. His behavior, style and looks may encourage or discourage certain replies from respondents. Careless mechanical processing may distort the findings. Errors may also creep in because of incorrect coding, faulty tabulation and/or statistical calculations, particularly in the data-analysis stage.
- ☆ **Instrument:** Error may arise because of the defective measuring instrument. The use of complex words, beyond the comprehension of the respondent, ambiguous meanings, poor printing, inadequate space for replies, response choice omissions, *etc.* are a few things that make the measuring instrument defective and may result in measurement errors. Another type of instrument deficiency is the poor sampling of the universe of items of concern. Researcher must know that correct measurement depends on successfully meeting all of the problems listed above. He must, to the extent possible, try to eliminate, neutralize or otherwise deal with all the possible sources of error so that the final results may not be contaminated.

Technique of Developing Measurement Tools

The technique of developing measurement tools involves a four-stage process, consisting of the following:

- ☆ Concept development;
- ☆ Specification of concept dimensions;
- ☆ Selection of indicators; and
- ☆ Formation of index.

The first and foremost step is that of concept development which means that the researcher should arrive at an understanding of the major concepts pertaining to his study. This step of concept development is more apparent in theoretical studies than in the more pragmatic research, where the fundamental concepts are often already established.

The second step requires the researcher to specify the dimensions of the concepts that he developed in the first stage. This task may either be accomplished by deduction *i.e.*, by adopting a more or less intuitive approach or by empirical correlation of the individual dimensions with the total concept and/or the other concepts. For instance, one may think of several dimensions such as product reputation, customer treatment, corporate leadership, concern for individuals, and sense of social responsibility and so forth when one is thinking about the image of a certain company. Once the dimensions of a concept have been specified, the researcher must develop indicators for measuring each concept element. Indicators are specific questions, scales, or other devices by which respondent's knowledge, opinion, expectation, *etc.*, are measured. As there is seldom a perfect measure of a concept, the researcher should consider several alternatives for the purpose. The use of more than one indicator gives stability to the scores and it also improves their validity. The last step is that of combining the various indicators into an index, *i.e.*, formation of an index. When we have several dimensions of a concept or different measurements of a dimension, we may need to combine them into a single index. One simple way for getting an overall index is to provide scale values to the responses and then sum up the corresponding scores. Such an overall index would provide a better measurement tool than a single indicator because of the fact that an "individual indicator has only a probability relation to what we really want to know." This way we must obtain an overall index for the various concepts concerning the research study.

Scaling

In research we quite often face measurement problem (since we want a valid measurement but may not obtain it), especially when the concepts to be measured are complex and abstract and we do not possess the standardized measurement tools. Alternatively, we can say that while measuring attitudes and opinions, we face the problem of their valid measurement. Similar problem may be faced by a researcher, of course in a lesser degree, while measuring physical or institutional concepts. As such we should study some procedures which may enable us to measure abstract concepts more accurately. This brings us to the study of scaling techniques.

Meaning of Scaling

Scaling describes the procedures of assigning numbers to various degrees of opinion, attitude and other concepts. This can be done in two ways *viz.* (i) making a judgment about some characteristic of an individual and then placing him directly on a scale that has been defined in terms of that characteristic and (ii) constructing questionnaires in such a way that the score of individual's responses assigns him a place on a scale. It may be stated here that a scale is a continuum, consisting of the highest point (in terms of some characteristic *e.g.*, preference, favorableness, *etc.*) and the lowest point along with several intermediate points between these two extreme points. These scale- point positions are so related to each other that when the first point happens to be the highest point, the second point indicates a higher degree in terms of a given characteristic as compared to the third point and the third point indicates a higher degree as compared to the fourth and so on. Numbers for measuring the distinctions of degree in the attitudes/opinions are, thus, assigned to individuals corresponding to their scale-positions. All this is better understood when we talk about scaling technique(s). Hence the term 'scaling' is applied to the procedures for attempting to determine quantitative measures of subjective abstract concepts. Scaling has been defined as a "procedure for the assignment of numbers (or other symbols) to a property of objects in order to impart some of the characteristics of numbers to the properties in question."

Scale Construction Techniques

In social science studies, while measuring attitudes of the people we generally follow the technique of preparing the opinionnaire* (or attitude scale) in such a way that the score of the individual responses assigns him a place on a scale. Under this approach, the respondent expresses his agreement or disagreement with a number of statements relevant to the issue. While developing such statements, the researcher must note the following two points:

- That the statements must elicit responses which are psychologically related to the attitude being measured;
- That the statements need be such that they discriminate not merely between extremes of attitude but also among individuals who differ slightly.

Researchers must as well be aware that inferring attitude from what has been recorded in opinionnaires has several limitations. People may conceal their attitudes and express socially acceptable opinions. They may not really know how they feel about a social issue. People may be unaware of their attitude about an abstract situation; until confronted with a real situation, they may be unable to predict their reaction. Even behavior itself is at times not a true indication of attitude. For instance, when politicians kiss babies, their behavior may not be a true expression of affection toward infants.

Thus, there is no sure method of measuring attitude; we only try to measure the expressed opinion and then draw inferences from it about people's real feelings or attitudes. With all these limitations in mind, psychologists and sociologists have developed several scale construction techniques for the purpose. The researcher should know these techniques so as to develop an appropriate scale for his own study. Some of the important approaches, along with the corresponding scales developed under each approach to measure attitude are asfollows:

Different Scales for Measuring Attitudes of People

Name of the scale construction approach	*Name of the scale developed*
1. Arbitrary approach	Arbitrary scales
2. Consensus scale approach	Differential scales (such as Thurstone Differential scale)
3. Item analysis approach	Summated scales (such as Likert Scale)
4. Cumulative scale approach	Cumulative scales (such as Guttman's Scalogram)
5. Factor analysis approach	Factor scales (such as Osgood's Semantic Differential, Multi-dimensional Scaling, etc.)

Arbitrary Scales

Arbitrary scales are developed on ad hoc basis and are designed largely through the researcher's own subjective selection of items. The researcher first collects few statements or items which he believes are unambiguous and appropriate to a given topic. Some of these are selected for inclusion in the measuring instrument and then people are asked to check in a list the statements with which they agree. The chief merit of such scales is that they can be developed very easily, quickly and with relatively less expense. They can also be designed to be highly specific and adequate. Because of these benefits, such scales are widely used in practice.

Differential Scales (or Thurston-type Scales)

The name of L.L. Thurston is associated with differential scales which have been developed using consensus scale approach. Under such an approach the selection of items is made by a panel of judges who evaluate the items in terms of whether they are relevant to the topic area and unambiguous in implication. The detailed procedure is as under:

- The researcher gathers a large number of statements, usually twenty or more, that express various points of view toward a group, institution, idea, or practice (*i.e.*, statements belonging to the topic area).
- These statements are then submitted to a panel of judges, each of whom arranges them in eleven groups or piles ranging from one extreme to another in position. Each of the judges is requested to place generally in the first pile the statements which he thinks are most unfavorable to the issue, in the second pile to place those statements which he thinks are next most unfavorable and he goes on doing so in this manner till in the eleventh pile he puts the statements which he considers to be the most favorable.

- ☆ This sorting by each judge yields a composite position for each of the items. In case of marked disagreement between the judges in assigning a position to an item, that item is discarded.
- ☆ For items that are retained, each is given its median scale value between one and eleven as established by the panel. In other words, the scale value of any one statement is computed as the 'median' position to which it is assigned by the group of judges.
- ☆ A final selection of statements is then made. For this purpose a sample of statements, whose median scores are spread evenly from one extreme to the other is taken. The statements so selected, constitute the final scale to be administered to respondents. The position of each statement on the scale is the same as determined by the judges.

After developing the scale as stated above, the respondents are asked during the administration of the scale to check the statements with which they agree. The median value of the statements that they check is worked out and this establishes their score or quantifies their opinion. It may be noted that in the actual instrument the statements are arranged in random order of scale value. If the values are valid and if the opinionnaire deals with only one attitude dimension, the typical respondent will choose one or several contiguous items (in terms of scale values) to reflect his views. However, at times divergence may occur when a statement appears to tap a different attitude dimension.

The Thurstone method has been widely used for developing differential scales which are utilized to measure attitudes towards varied issues like war, religion, *etc.* Such scales are considered most appropriate and reliable when used for measuring a single attitude. But an important deterrent to their use is the cost and effort required to develop them. Another weakness of such scales is that the values assigned to various statements by the judges may reflect their own attitudes. The method is not completely objective; it involves ultimately subjective decision process. Critics of this method also opine that some other scale designs give more information about the respondent's attitude in comparison to differential scales.

Summated Scales (or Likert-type Scales)

Summated scales (or Likert-type scales) are developed by utilizing the item analysis approach wherein a particular item is evaluated on the basis of how well it discriminates between those persons whose total score is high and those whose score is low. Those items or statements that best meet this sort of discrimination test are included in the final instrument.

Thus, summated scales consist of a number of statements which express either a favorable or unfavorable attitude towards the given object to which the respondent is asked to react. The respondent indicates his agreement or disagreement with each statement in the instrument. Each response is given a numerical score, indicating its favorableness or unfavorableness, and the scores are totalled to measure the respondent's attitude. In other words, the overall score represents the respondent's position on the continuum of favorable-unfavorableness towards an issue.

Most frequently used summated scales in the study of social attitudes follow the pattern devised by Likert. For this reason they are often referred to as Likert-type scales. In a Likert scale, the respondent is asked to respond to each of the statements in terms of several degrees, usually five degrees (but at times 3 or 7 may also be used) of agreement or disagreement. For example, when asked to express opinion whether one considers his job quite pleasant, the respondent may respond in any one of the following ways:

i. Strongly agree,
ii. Agree,
iii. Undecided,
iv. Disagree,
v. Strongly disagree.

We find that these five points constitute the scale. At one extreme of the scale there is strong agreement with the given statement and at the other, strong disagreement, and between them lie intermediate points. We may illustrate this as under:

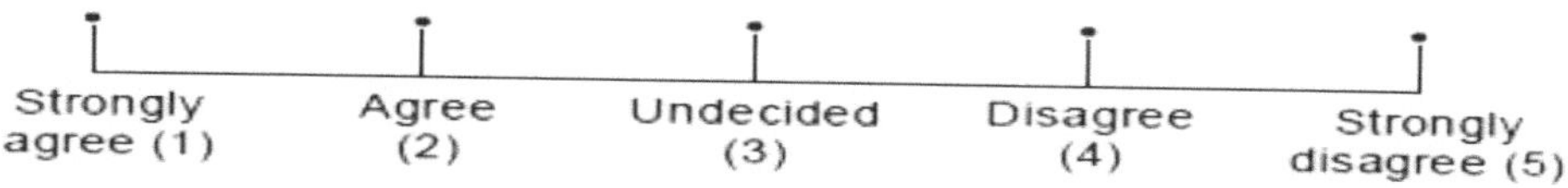

Each point on the scale carries a score. Response indicating the least favorable degree of job satisfaction is given the least score (say 1) and the most favorable is given the highest score (say 5). These score—values are normally not printed on the instrument but are shown here just to indicate the scoring pattern. The Likert scaling technique, thus, assigns a scale value to each of the five responses. The same thing is done in respect of each and every statement in the instrument. This way the instrument yields a total score for each respondent, which would then measure the respondent's favorableness toward the given point of view. If the instrument consists of, say 30 statements, the following score values would berevealing.

30 × 5 = 150 Most favorable response possible

30 × 3 = 90 A neutral attitude

30 × 1 = 30 Most unfavorable attitude.

The scores for any individual would fall between 30 and 150. If the score happens to be above 90, it shows favorable opinion to the given point of view, a score of below 90 would mean unfavorable opinion and a score of exactly 90 would be suggestive of a neutral attitude.

Procedure

The procedure for developing a Likert-type scale is as follows:

1. As a first step, the researcher collects a large number of statements which are relevant to the attitude being studied and each of the statements expresses definite favorableness or unfavorableness to a particular point of view or the attitude and that the number of favorable and unfavorable statements is approximately equal.
2. After the statements have been gathered, a trial test should be administered to a number of subjects. In other words, a small group of people, from those who are going to be studied finally, are asked to indicate their response to each statement by checking one of the categories of agreement or disagreement using a five point scale as stated above.
3. The response to various statements are scored in such a way that a response indicative of the most favorable attitude is given the highest score of 5 and that with the most unfavorable attitude is given the lowest score, say, of1.
4. Then the total score of each respondent is obtained by adding his scores that he received for separate statements.
5. The next step is to array these total scores and find out those statements which have a high discriminatory power. For this purpose, the researcher may select some part of the highest and the lowest total scores say the top 25 per cent and the bottom 25 per cent. These two extreme groups are interpreted to represent the most favorable and the least favorable attitudes and are used as criterion groups by which to evaluate individual statements. This way we determine which statements consistently correlate with low favorability and which with high favorability.
6. Only those statements that correlate with the total test should be retained in the final instrument and all others must be discarded from it.

Advantages

The Likert-type scale has several advantages. Mention may be made of the important ones.

- ☆ It is relatively easy to construct the Likert-type scale in comparison to Thurstone-type scale because Likert-type scale can be performed without a panel of judges.
- ☆ Likert-type scale is considered more reliable because under it respondents answer each statement included in the instrument. As such it also provides more information and data than does the Thurstone-typescale.

- ☆ Each statement, included in the Likert-type scale, is given an empirical test for discriminating ability and as such, unlike Thurstone-type scale, the Likert-type scale permits the use of statements that are not manifestly related (to have a direct relationship) to the attitude being studied.
- ☆ Likert-type scale can easily be used in respondent-centered and stimulus- centered studies *i.e.*, through it we can study how responses differ between people and how responses differ between stimuli.
- ☆ Likert-type scale takes much less time to construct; it is frequently used by the students of opinion research. Moreover, it has been reported in various research studies* that there is high degree of correlation between Likert-type scale and Thurstone-type scale.

Limitations

There are several limitations of the Likert-type scale as well. One important limitation is that, with this scale, we can simply examine whether respondents are more or less favorable to a topic, but we cannot tell how much more or less they are. There is no basis for belief that the five positions indicated on the scale are equally spaced. The interval between 'strongly agree' and 'agree', may not be equal to the interval between "agree" and "undecided". This means that Likert scale does not rise to a stature more than that of an ordinal scale, whereas the designers of Thurstone scale claim the Thurstone scale to be an interval scale. One further disadvantage is that often the total score of an individual respondent has little clear meaning since a given total score can be secured by a variety of answer patterns. It is unlikely that the respondent can validly react to a short statement on a printed form in the absence of real-life qualifying situations. Moreover, there "remains a possibility that people may answer according to what they think they should feel rather than how they do feel." In spite of all the limitations, the Likert-type summated scales are regarded as the most useful in a situation wherein it is possible to compare the respondent's score with a distribution of scores from some well-defined group. They are equally useful when we are concerned with a programme of change or improvement in which case we can use the scales to measure attitudes before and after the programme of change or improvement in order to assess whether our efforts have had the desired effects.

We can as well correlate scores on the scale to other measures without any concern for the absolute value of what is favorable and what is unfavorable. All this accounts for the popularity of Likert-type scales in social studies relating to measuring of attitudes.

Cumulative Scales

Cumulative scales or Louis Guttman's scalogram analysis, like other scales, consists of series of statements to which a respondent expresses his agreement or disagreement. The special feature of this type of scale is that statements in it form a cumulative series. This, in other words, means that the statements are related to one another in such a way that an individual, who replies favorably to say item No. 3, also replies favorably to items No. 2 and 1, and one who replies favorably to item No. 4 also replies favorably to items No. 3, 2 and 1, and so on. This being so an individual whose attitude is at a certain point in a cumulative scale will answer favorably all the items on one side of this point, and answer unfavorably all the items on the other side of this point. The individual's score is worked out by counting the number of points concerning the number of statements he answers favorably. If one knows this total score, one can estimate as to how a respondent has answered individual statements constituting cumulative scales. The major scale of this type of cumulative scales is the Guttman's scalogram. We attempt a brief description of the same below.

The technique developed by Louis Guttman is known as scalogram analysis, or at times simply 'scale analyses. Scalogram analysis refers to the procedure for determining whether a set of items forms a unidimensional scale. A scale is said to be unidimensional if the responses fall into a pattern in which endorsement of the item reflecting the extreme position results also in endorsing all items which are less extreme. Under this technique, the respondents are asked to indicate in respect of each item whether they agree or disagree with it, and if these items form a unidimensional scale, the response pattern will be as under:

Response Pattern in Scalogram Analysis

Item Number				*Respondent Score*
4	*3*	*2*	*1*	
X	X	X	X	4
–	X	X	X	3
–	–	X	X	2
–	–	–	X	1
–	–	–	–	0

X = Agree

– = Disagree

A score of 4 means that the respondent is in agreement with all the statements, which is indicative of the most favorable attitude. But a score of 3 would mean that the respondent is not agreeable to item 4, but he agrees with all others. In the same way one can interpret other values of the respondents' scores. This pattern reveals that the universe of content is scalable.

Procedure

The procedure for developing a scalogram can be outlined as under:

- The universe of content must be defined first of all. In other words, we must lay down in clear terms the issue we want to deal within our study.
- The next step is to develop a number of items relating the issue and to eliminate by inspection the items that are ambiguous, irrelevant or those that happen to be too extreme items.
- The third step consists in pre-testing the items to determine whether the issue at hand is scalable (The pretest, as suggested by Guttman, should include 12 or more items, while the final scale may have only 4 to 6 items. Similarly, the number of respondents in a pretest may be small, say 20 or 25 but final scale should involve relatively more respondents, say 100 or more). In a pretest the respondents are asked to record their opinions on all selected items using a Likert-type 5-point scale, ranging from 'strongly agree' to 'strongly disagree'. The strongest favorable response is scored as 5, whereas the strongest unfavorable response as 1. The total score can thus range, if there are 15 items in all, from 75 for most favorable to 15 for the least favorable. Respondent opinionnaires are then arrayed according to total score for analysis and evaluation. If the responses of an item form a cumulative scale, its response category scores should decrease in an orderly fashion as indicated in the above table. Failure to show the said decreasing pattern means that there is overlapping which shows that the item concerned is not a good cumulative scale item *i.e.*, the item has more than one meaning. Sometimes the overlapping in category responses can be reduced by combining categories. After analyzing the pretest results, a few items, say 5 items, may be chosen.
- The next step is again to total the scores for the various opinionnaires, and to rearray them to reflect any shift in order, resulting from reducing the items, say, from 15 in pretest to, say, 5 for the final scale. The final pretest results may be tabulated in the form of a table given in below.

The Final Pretest Results in a Scalogram Analysis

Scale type			*Item*			*Errors*	*Number of*	*Number of*
	5	*12*	*3*	*10*	*7*	*per case*	*cases*	*errors*
5 (perfect)	X	X	X	X	X	0	7	0
4 (perfect)	–	X	X	X	X	0	3	0
(nonscale)	–	X	–	X	X	1	1	1
(nonscale)	–	X	X	–	X	1	2	2
3 (perfect)	–	–	X	X	X	0	5	0
2 (perfect)	–	–	–	X	X	0	2	0
1 (perfect)	–	–	–	–	X	0	1	0
(nonscale)	–	–	X	–	–	2	1	2
(nonscale)	–	–	X	–	–	2	1	2
0 (perfect)	–	–	–	–	–	0	2	0
			$n = 5$				$N = 25$	$e = 7$

The table shows that five items (numbering 5, 12, 3, 10 and 7) have been selected for the final scale. The number of respondents is 25 whose responses to various items have been tabulated along with the number of errors. Perfect scale types are those in which the respondent's answers fit the pattern that would be reproduced by using the person's total score as a guide. Non-scale types are those in which the category pattern differs from that expected from the respondent's total score *i.e.*, non-scale cases have deviations from unidimensionality or errors. Whether the items (or series of statements) selected for final scale may be regarded a perfect cumulative (or a unidimensional scale), we have to examine on the basis of the coefficient of reproducibility. Guttman has set 0.9 as the level of minimum reproducibility in order to say that the scale meets the test of unidimensionality. He has given the following formula for measuring the level of reproducibility:

Guttman's Coefficient of Reproducibility = $1 - e/n(N)$

where,

e = number of errors

n = number of items

N = the number above of table cases figures.

Coefficient of Reproducibility = $1 - 7/5(25) = .94$

This shows that items number 5, 12, 3, 10 and 7 in this order constitute the cumulative or unidimensional scale, and with this we can reproduce the responses to each item, knowing only the total score of the respondent concerned. Scalogram, analysis, like any other scaling technique, has several advantages as well as limitations. One advantage is that it assures that only a single dimension of attitude is being measured. Researcher's subjective judgment is not allowed to creep in the development of scale since the scale is determined by the replies of respondents. Then, we require only a small number of items that make such a scale easy to administer. Scalogram analysis can appropriately be used for personal, telephone or mail surveys. The main difficulty in using this scaling technique is that in practice perfect cumulative or unidimensional scales are very rarely found and we have only to use its approximation testing it through coefficient of reproducibility or examining it on the basis of some other criteria. This method is not a frequently used method for the simple reason that its development procedure is tedious and complex. Such scales hardly constitute a reliable basis for assessing attitudes of persons towards complex objects for predicting the behavioral responses of individuals towards such objects. Conceptually, this analysis is a bit more difficult in comparison to other scaling methods.

Factor Scales

Factor scales are developed through factor analysis or on the basis of inter correlations of items which indicate that a common factor accounts for the relationships between items. Factor scales are particularly "useful in uncovering latent attitude dimensions and approach scaling through the concept of multiple-dimension attribute space." More specifically,the two problems *viz.*, how to deal appropriately with the universe of content which is multi-dimensional and how to uncover underlying (latent) dimensions which have not been identified, are dealt with through factor scales. An important factor scale based on factor analysis is Semantic Differential (S.D.) and the other one is Multidimensional Scaling. We give below a brief account of these factor scales.

Semantic differential scale: Semantic differential scale or the S.D. scale developed by Charles E. Osgood, G.J. Suci and P.H. Tannenbaum (1957), is an attempt to measure the psychological meanings of an object to an individual. This scale is based on the presumption that an object can have different dimensions of connotative meanings which can be located in multidimensional property space, or what can be called the semantic space in the context of S.D. scale. This scaling consists of a set of bipolar rating scales, usually of 7 points, by which one or more respondent's rate one or more concepts on each scale item. For instance, the S.D. scale items for analyzing candidates for leadership position may be shown as under:

		3	2	1	0	–1	–2	–3	
(E)	Successful								Unsuccessful
(P)	Severe								Lenient
(P)	Heavy								Light
(A)	Hot								Cold
(E)	Progressive								Regressive
(P)	Strong								Weak
(A)	Active								Passive
(A)	Fast								Slow
(E)	True								False
(E)	Sociable								Unsociable

Candidates for leadership position (along with the concept—the 'ideal' candidate) may be compared and we may score them from +3 to –3 on the basis of the above stated scales. (The letters, E, P, A showing the relevant factor *viz.*, evaluation, potency and activity respectively, written along the left side are not written in actual scale. Similarly the numeric values shown are also not written in actual scale.) Osgood and others did produce a list of some adjective pairs for attitude research purposes and concluded that semantic space is multidimensional rather than unidimensional. They made sincere efforts and ultimately found that three factors, *viz.*, evaluation, potency and activity, contributed most to meaningful judgments by respondents. The evaluation dimension generally accounts for 1/2 and 3/4 of the extractable variance and the other two factors account for the balance.

Procedure

Various steps involved in developing S.D. scale are as follows:

- ☆ First of all the concepts to be studied are selected. The concepts are usually chosen by personal judgment, keeping in view the nature of the problem.
- ☆ The next step is to select the scales bearing in mind the criterion of factor composition and the criterion of scale's relevance to the concepts being judged (it is common practice to use at least three scales for each factor with the help of which an average factor score has to be worked out). One more criterion to be kept in view is that scales should be stable across subjects and concepts.
- ☆ Then a panel of judges are used to rate the various stimuli (or objects) on the various selected scales and the responses of all judges would then be combined to determine the composite scaling. To conclude, "the S.D. has a number of specific advantages. It is an efficient and easy way to secure attitudes from a large sample. These attitudes may be measured in both direction and intensity. The total set of responses provides a comprehensive picture of the meaning of an object, as well as a measure of the subject doing the rating. It is a standardized technique that is easily repeated, but escapes many of the problems of response distortion found with more direct methods."

Multidimensional Scaling

Multidimensional scaling (MDS) is relatively more complicated scaling device, but with this sort of scaling one can scale objects, individuals or both with a minimum of information. Multidimensional scaling (or MDS) can be characterized as a set of procedures for portraying perceptual or affective dimensions of substantive interest. It "provides useful methodology for portraying subjective judgments of diverse kinds." MDS is used when all the variables (whether metric or non-metric) in a study are to be analyzed simultaneously and all such variables happen to be independent. The underlying assumption in MDS is that people (respondents) "perceive a set of objects as being more or less similar to one another on a number of dimensions (usually uncorrelated with one another) instead of only one." Through MDS techniques one can represent geometrically the locations and interrelationships among a set of points. In fact, these techniques attempt to locate the points, given the information about a set of interpoint distances, in space of one or more dimensions such as to best summarize the information contained in the inter point distances. The distances in the solution space then optimally reflect the distances contained in the input data. For instance, if objects, say X and Y, are thought of by the respondent as being most similar as compared to all other possible pairs of objects, MDS techniques will position objects X and Y in such a way that the distance between them in multidimensional space is shorter than that between any two other objects.

Two approaches, *viz.*, the metric approach and the non-metric approach, are usually talked about in the context of MDS, while attempting to construct a space containing m points such that m(m – 1)/2 inter point distances reflect the input data. The metric approach to MDS treats the input data as interval scale data and solves applying statistical methods for the additive constant* which minimizes the dimensionality of the solution space. This approach utilizes all the information in the data in obtaining a solution. The data (*i.e.*, the metric similarities of the objects) are often obtained on a bipolar similarity scale on which pairs of objects are rated one at a time. If the data reflect exact distances between real objects in an r-dimensional space, their solution will reproduce the set of inter point distances. But as the true and real data are rarely available, we require random and systematic procedures for obtaining a solution. Generally, the judged similarities among a set of objects are statistically transformed into distances by placing those objects in a multidimensional space of some dimensionality.

The non-metric approach first gathers the non-metric similarities by asking respondents to rank order all possible pairs that can be obtained from a set of objects. Such non-metric data is then transformed into some arbitrary metric space and then the solution is obtained by reducing the dimensionality. In other words, this non- metric approach seeks "a representation of points in a space of minimum dimensionality such that the rank order of the inter point distances in the solution space maximally corresponds to that of the data. This is achieved by requiring only that the distances in the solution be monotone with the input data."9 The non-metric approach has come into prominence during the sixties with the coming into existence of high speed computers to generate metric solutions for ordinal input data. The significance of MDS lies in the fact that it enables the researcher to study "the perceptual structure of a set of stimuli and the cognitive processes underlying the development of this structure. Psychologists, for example, employ multidimensional scaling techniques in an effort to scale psychophysical stimuli and to determine appropriate labels for the dimensions along which these stimuli vary." The MDS techniques, in fact, do away with the need in the data collection process to specify the attribute(s) along which the several brands, say of a particular product, may be compared as ultimately the MDS analysis itself reveals such attribute(s) that presumably underlie the expressed relative similarities among objects. Thus, MDS is an important tool in attitude measurement and the techniques falling under MDS promise "a great advance from a series of one-dimensional measurements (*e.g.*, a distribution of intensities of feeling towards single attribute such as color, taste or a preference ranking with indeterminate intervals), to a perceptual mapping in multidimensional space of objects. company images, advertisement brands, *etc.*" In spite of all the merits stated above, the MDS is not widely used because of the computation complications involved under it. Many of its methods are quite laborious in terms of both the collection of data and the subsequent analyses. However, some progress has been achieved (due to the pioneering efforts of Paul Green and his associates) during the last few years in the use of non-metric MDS in the context of market research problems. The techniques have been specifically applied in "finding out the perceptual dimensions, and the spacing of stimuli along these dimensions, that people, use in making judgments about the relative similarity of pairs of Stimuli." But, "in the long run, the worth of MDS will be determined by the extent to which it advances the behavioral sciences."

— Chapter 7 —

Participatory Approaches in Community Based Observation (CBD) and Researches

7.1 People's Participation in Observation, Analysis and Planning for Rural Development

Why Should People Participate?

There is no controversy as to whether there should be people's participation in the development programmes in the development programmes designed to meet the basic needs of the people. The post-mortem of any unsuccessful programmes has shown that they have invariably failed when they failed to provide an adequate place and importance to people's participation in the planning and analysis.

There is a body of philosophy and a set of assumptions behind people's participation in rural development. These are;

i) It should be people's programme with government's participation;

ii) Community needs to build up their own capacity, agents have got a catalytic role;
iii) Self- imposed changes have got permanence as compared to those imposed from outsides
iv) Holistic approaches are far better than the fragmented approaches;
v) People need certain help to solve unique problems;
vi) Barriers that prevent active participation, inhibit personal development as well;
vii) Given the opportunity, people can innovate solution to their problems.
viii) Apparently most helpless community can be an active factor in helping themselves by dint of their extreme wisdom.

What is Participatory Rural Appraisal?

Participatory Rural Appraisal is (PRA) is a methodology which helps in interacting with local communities, understanding them and learning from them. It helps in the process of involvement with local communities for indigenous knowledge-building exercises. It is a way of learning from and with community members to investigate, analyze and evaluate constraints and opportunities and make informed and timely decisions regarding development projects.

It involves a set of principles, a process of communication and a menu of techniques for seeking villagers' participation in putting forward their points of view about any issue and enabling them do their own analysis and offer their own contribution in different forms with a view to make use of such learning. It initiates a participatory process and sustains it. Its principles and the menu of techniques help in organizing participation.

PRA is a means of generating different kinds of data, identifying and mobilizing intended groups and evoking their participation and also opening ways in which intended groups can participate in decision making, project design, execution, monitoring and evaluation. It provides an alternative framework for data collection and analysis. Because of its participatory nature, it is an useful methodology to focus attention on people, their livelihood and their inter-relationship with socio-economic and ecological factors.

PRA is sometimes known as Rapid Rural Appraisal (RRA)/Participatory Rapid Appraisal (PRA), where the emphasis is on both 'rapid and participatory'. The emphasis on 'rapid' however, is more in terms of data collection and less in terms of process of development or even implementation of plans. Other names are also used for describing PRA, some of which are Participatory Rapid Rural Appraisal) and Participatory Learning Method (PALM). PRA, as Practiced in the field has given rise to different schools of PRA with difference in style and emphasis (Chambers: 1991).

7.2 Indigenous Criteria of Wealth and Poverty

The villagers seemed to have a mental checklist which they were referring to for grouping of households, the checklist for determining their criteria for well-being. They identified poor households on the basis of the conditions and causes of poverty prevailing in such households. Food availability including wild food, during a year was the most significant factor for the villagers in grouping of poor households. The size of landholdings and its productivity was another important consideration for the villagers for identifying higher groups but land alone was not enough. Its productive capacity was equally important. The number of dependent was also another consideration apart from food availability and land holdings.

On the basis of semi-structured interviews the reason for poverty and deprivation were highlighted by the villagers. Environmental degradation had accentuated the uncertainty of food availability from common property resources and increased the incidence of poverty. Common property resources lent support in several ways. For many, such resources formed the basis of their livelihood, for other it supplemented income, food and other requirements and also provided sustenance during difficult time. Environmental degradation had reduced the quality of life for the poor. It had let to increased deprivation. The poor are no longer able to fall back on the environment for food, water, work, medicine, fuel wood and clean air especially during lean periods and in years of crop failures. The degradation of pond pastures, riverbeds, decline in the number of trees and shrubs, decline in area, decline in quality of water and the erosion of forests as source of food, fodder and fuel wood had directly affected their sources of livelihood, cost of living, its standards, costs and health.

7.3 Comparative Analysis: Paradigms, Principles, Objectives and Evaluation

It is important that analysis of evidence and experience concerning PRA- type enquiry and questioner survey is preceded by a discussion of the foundations of the two systems of enquiry under review. This would help in understanding and appreciating the underlying differences, if any, of the two systems in the context of their paradigmatic bases, principles, objectives and way of evaluation. This would also constitute a base of for reviewing field experience under the two systems of enquiry.

Any enquiry whether through questionnaire survey or PRA would be guided by their respective belief systems which are known as paradigms. Each paradigm responds to three basic questions concerning the nature of reality (ontology), the nature of relationship between knower and the Known (epistemology) and the methodology involved (Guba: 1990). In this chapter I **examine the two systems of inquiry,** the questionnaire survey and the PRA Method on the basis the following criteria:

- ✰ Paradigmatic Basis
- ✰ Principles
- ✰ Objectives
- ✰ Evaluation of Data

Paradigmatic Basis and Positivism

The methodologies of questionnaire survey and PRA emerge from different paradigms. The differences in their paradigmatic bases are a key to understanding the diversities in the nature of systems of inquiry and their practice. To quote Schwandt (1990)," to Study a methodology is not simply to examine the exercise of method, it is to study a way of knowing...... to explore a logic of justification or a meta framework for understanding the exercise of method, that is for examining the principles and procedures by which we formulate inquiry problems, and evaluate the correctness and profundity of those answers".

Positivism constitutes the basic belief system of conventional inquiry in which lies the paradigmatic basis of the questionnaire –survey. The nature of reality of positivism constitutes reality which exists 'out there' and is driven by immutable natural laws and mechanisms. Its nature of relationship between the knower or inquirer and the known is that of an objectivist. It is both possible and essential for the inquirer to adopt an objective approach to the inquiry.

The methodology of positivism is scientific. Question and hypothesis are posed in advance in forms of propositions and are subjected to empirical testing under carefully controlled conditions. To state it in simple words, the scientific methodology is principally concerned with procedures for the development and testing of causal hypothesis. The corner stone of the methodology lie in observation, measurement experiment and theory building with statistics playing a major role in formulation and testing of hypothesis.

In social disciplines, our understanding of methodology has been generally based on the conception of scientific inquiry, the principles and procedures that govern investigations of the physical world. In line with this, our professional belief in the naturalistic interpretation of social science is also pronounced. We regard the scientific methodology as a sure and an independent way of being rational in our inquiry.

Paradigm Blocked and Post Positivism

Any paradigm can block the level of inquiry in four ways:

- ✰ Absolute assertion
- ✰ The claim that certain things can never be known
- ✰ The claim that certain elements of science are basic and independent of all else

- The claim that a law or truth has found its last and perfect formulation.
- In the emerging era of post positivism, such claims positivism are being increasingly challenged in social disciplines. The methodology of natural science is being questioned and being replaced by methods drawn from humanistic disciplines. In this context, post-positivism recognizes that many imbalance of positivism have been allowed in the name of achieving realistic and objective inquiry. Some of them are as follows:
- The imbalance between rigor and relevance, where, search for more rigor to establish internal validity make the findings less generalized. Thus the laboratory results are generalizable only to another laboratory.
- The imbalance between precision and richness which leads to over emphasis or quantitative methods and precise and accurate responses.
- The imbalance between discovery and verification which leads to acceptability of selected results since there are not many objective ways of verifying all kind of results.
- The imbalance between elegance and applicability where the emphasis is on the statement of formal and reductionist theories and their testing and applications. Such theories may not be appropriate in the local context thus making elegant theories redundant and unrealistic.

The Constructivist Paradigm

The history of science and inquiry reveals that our knowledge of both the external world and the ways and means of human interaction are always evolving (Schwandt, 1990). Hence some paradigms become old while others are relatively new and emerging. One such recent paradigm is that of constructivist paradigm. Methodologies of such paradigm aim at understanding as much as possible the following:

- Aspects of human experiences as it is lives or felt or undergone as participants in that experience
- The Context of that experience
- The naturally occurring context
- Procedures for considering the context and experience as a complex, temporal, socio-cultural and geographic whole
- The inquirer employing ordinary field work methods.

The constructivist methodologies reject the naturalistic interpretation of the social sciences and aim at inquiring and interpreting the realm of inter-subjective meanings as constituted in culture, language, signs, symbol *etc.* One contribution to systems of enquiry from the constructivist paradigm is the emergence of participatory inquiry whose framework of analysis is constructive rather than naturalistic.

The Paradigm of 'Critical Realism'

The paradigm of 'critical realism' is another devilment in the era of positivism from which participatory inquiry has routed its evaluation of findings. The nature of reality is such that the enquirer needs to be critical about findings because of human initiations regarding knowledge of reality. Critical realism recognize that findings emerge from the interaction of inquirer inquired and hence proposes a modified objectivity by relaying on 'critical tradition' which implies that report on any inquiry needs to be consistent with the existing tradition of the field and subjects every inquiry to peer group judgment in the 'critical community' which implies that report on any inquiry needs to be consistent with the existing tradition of the field and subjects every inquiry to peer group judgment in the 'critical community'. This makes it difficult for old and new paradigm to assert them.

Methodologically, 'critical realism generally looks for critical multiplicity which is elaborated triangulation. According to the paradigm of critical realism, it is essential that the 'findings' of an enquiry be based on 'triangulation' of as many sources ... of data, investigators, theories and methods, place and time. as possible because of inherent human limitations of knowledge, limitation in detecting errors and overcoming biases.

Participatory Inquiry and Post-positivism

Participatory Rural Appraisal (PRA) is based on post positivism to combine the strength of 'constructivist paradigm' and that of 'critical realism'. From the constructivist paradigm, PRA type of inquiry seeks to understand human experience as it is lived and perceived by the participants, the context of the experience add an interdisciplinary framework for analysis of the experience. Critical realism helps PRA in evaluating such analysis and correcting for errors and biases through multiple sources, methods, locations and other ways.

All paradigms are human constructions and hence subject to biases and errors. With post positivism we recognize that any methodology is historically situated and it evolves overtime. Our methodologies for inquiry for inquiry and our standards for judging those methodologies are being continually revised by a process of mutual adjustment (Guba: 1990).

7.4 The Principles Governing Questionnaire Survey and PRA

The principles governing the methods are quite different. Conventional methods such as questionnaire survey primarily seek a general lawwhich helps in explaining special event under given conditions. Based on a conceptual framework, the general law which helps in explaining a special event under given conditions. Based on a conceptual framework, the general law operates within given conditions. The framework shows the relationship amongst different variables, magnitudes of which can be measured by collecting data and testing it empirically according to such criteria as objectivity, reliability, replication and validity.

In case of a questionnaire survey we are essentially trying to know more about the universe either by deducing from a sample or by complete enumeration of the population in the universe. One is basically trying to find a general picture along certain lines which means a set of common questions are to be repeated in a questionnaire and investigations carried out carried out on that basis whether of a household or of an individual. Given the questions, the answers are stratified and classified into different groups (common clusters of responses are identified for classifying responses into different groups) to find out the patterns underlying those responses whether across time or across individuals, households and groups. The focus of surveys is on identifying general trends and patterns.

As compared to survey method, PRA is primarily for analysis of differences in local phenomena and processes. It is based on people's perceptions for analysis of complexities and heterogeneity underlying socio-economic systems and processes. The answer to 'averages' and set of patterns can be found only when analysis of difference has been conducted.

The Principles of PRA

The following gives an account of the principles guiding PRA. These principles help in improving the quality of participation, ensuring depth and breadth of information generated, the process involved and also provided a mechanism for evaluation.

i. Optimal Ignorance

In order to minimize cost and time, the principle of optimal ignorance is applied by investigators which mean knowing what is worth knowing and knowing enough to serve the purpose and not knowing the rest or not trying to find out more. Associated with this is seeking appropriate imprecision or avoiding precision that is not necessary (Chambers: 1993)

ii. Seeking Diversity

PRA is concerned more with analysis of difference rather than looking for representativeness of results or data collected. It is looking for diverse events, different processes and forces explaining various relationships in local communities.

iii. Offsetting Biases and Triangulating

PRA is essentially a methodology to offset biases of the questionnaire survey method. It aims offsetting such biases by being relaxed and not rushing, listening and not lecturing, probing and not speeding indifferently and looking for participation of poor and other weaker sections of local communities (Chambers, 1993).

The process of triangulation is an important part of PRA. For any data generated it is essential to check the reliability and validity of the data by putting them to different tests. Triangulation is cross-checking the data in diffe4rent ways. This is done through use of various methods and by using different ways to validate information. 'Triangulation is adopted as a principle to improve trustworthiness of data. It involves conscious, non-random selection in different dimensions such as (a) team composition, (b) Units of observation, (c) PRA techniques, (d) locations, (e) time and (f) sources.

iv. Listening and Learning, Learning Rapidly and Progressively and Learning through Participation

Local communities have their experiences, their history and culture, their ideas, their priorities and references. Listening to local people helps in portraying their points of view which otherwise remain unknown. The amount of learning can increase progressively with participation of those who form the subject of inquiry. In this context, the principle of openness operates where an investigator facilitates and individual's discussion and views, in a natural setting, so as to interact from within. PRA is also based on the principle of communication. Communication is not only an instrument to cull out expressions, giving information but also an instrument of clarification of the importance of those expressions from the speaker (Streiffeler: 1991)

The above principles contribute towards different features of PRA which promote participation and learning. Some of such features are that learning is iterative, methods are innovative and interactive and the process of PRA remains informal so as to facilitate participation.

A questionnaire survey, in principle, is looking for uniformity, identifying patterns and/or discovering general laws in an universe with regards to a sample population where the output of the survey and its validity and reliability are of major importance. Whereas, PRA is based on principles which emphasize the process of learning and analysis of difference and also to evaluate its findings. In PRA, the process of learning is as important as the output emerging from the process.

How Errors can Arise?

Errors can be of two kinds (i) Bias (which includes misunderstanding, misreading, lying, inventing answers, differential/prudent replies and others similar kind) is a systematic error and (ii) noise is a kind of error which is unsystematic and random.

Errors in data collection can be present in the form of bias which can have different sources. Bias in some form and for some reason is present in most data sets and there are several sources through which it can arise. It is a systematic error having an identifiable pattern and can be minimized in different ways. For factual questions it is possible to check the respondent's answer against are liable record. For questions about subjective phenomena such as opinions, feelings or perceptions there are not many direct ways to access the accuracy of answers.

The sources of error giving rise to bias are:

i. Expectations of the researcher;
ii. Such expectation as communicated to people *e.g.* wording of a question or even the position of questions can affect answers;
iii. Age, profession, sex religion, race, language, dress *etc.* of researcher;
iv. Behaviour of people and their characteristics;
v. Research plan design;
vi. Selection of area, place, time and people *e.g.* setting of an interview: and
vii. One event selected from a set of events whose results might change if other events are chosen.

As regards biases (i). and, (ii). Clues given by aresearcher can be such so as to have results in line with the expectation s of the researcher. This can lead to creation of bias thus defeating the very objective of an investigation. One way in which a questionnaire survey can take care of this bias is to keep both people and investigators uninformed about the purpose of the study so that biases from both sides do not affect the study. However, it may happen that the efficiency of the investigator falls in cases in which the investigator remains uninformed about such purpose. In case of PRA, since people's participation play an important role, participation of people and their analysis can guard against the expectations play an important role, participation of people and their analysis can guard against the expectations of the researchers. Or alternatively, the researcher is supposed to learn from the people's knowledge and views and is not expected to go with a pre-set mind and foregone conclusions.

Biases such as of category (iii) can originate through age, profession, sex and race of a researcher which can affect the respondent's answers. In case of a health survey directed towards ascertaining the percentage of women suffering from communicable diseases, respondents can either avoid male respondents or be reluctant to answer questions or give misleading answers. In case of bias arising through the profession of the researcher, it may be observed that each profession tends to carry it own bias and professional outlook of professional dependents on their training and experience. For instance, economists tend to see the economic aspects of rural problems while engineers are more concerned with the engineering aspects of a problem. It is difficult to find people who naturally have total perspective of a problem at hand.

In PRA exercise, the different sources of bias through age, sex and race can be balanced by altering composition of team members and mixing people of different age, sex and race. Hence, there exists considerable scope to take care of such a bias.

Bias arising through behaviour of people is indicated in category (iv) above. Those people sensitive to the expectations of a researcher can knowingly guess them, *e.g.* through leading questions and ceases to be their normal selves and move the analysis in line with the researcher's expectations. The framing of questions can be such that they are not neutral and the people are able to guess the expectations. The framing of questions can be such that they are not neutral and the people are able to guess the expectations of the researcher. For a PRA type of exercise, one way to eliminate such a bias is but not asking leading and closed questions.

7.5 Criteria for Trustworthiness of Data

The question is what data to accept and what to reject; how to reduce errors and to increase the trustworthiness of data. For this purpose, there can be two kinds of criteria which can be applied- the internal criteria and the external criteria. In simple term, the external criteria consist of all that the data has to say and to compare it with what is already known about the areas to which the data relates. The internal criterion, on the other hand, is concerned about the accuracy of the process and method used for collection of data. Both criteria are important for the evaluation of the data collected (Katzer, Cook and Crouch: 1982).

As far as the external criteria are concerned, not many methodologies have provision within them for a comparative evaluation of data already available on the subject. Much depends on the aptitude of the researcher to evaluate the data on the basis of studies available in that area. However, as far as PRA is concerned, secondary data review forms an important method through which background material is provided for a PRA exercise. It is considered as one of the techniques to help in better understanding of the subject concerned.

With regard to internal criteria, in the conventional approach, one can estimate an average error term and examine the deviations of the actual observations from the average of the expected observation. Greater the deviations greater are the imperfections in the data for such deviations do not help in forming near relationship amongst variables so as to explain a particular phenomenon. The parameters of the equation explaining the relationship would have low acceptability in standard statistical terms and low explanatory power to explain that relationship.

What is Triangulation?

Team composition in generally interdisciplinary and can be constituted in such a manner so as to have the information generated from different perspectives. Teams can have persons from different disciplines, sex, culture and experience so that data emerging from PRA can be discuss within a team and with local people by a team to clarify issues and select issue foe for further probing. The unit of observation can be changed in order to cross check the information gained one also has the option of going for deeper probing if necessary. Different groups can be approached separately or in combination so as to arrive at views from different angles. Other sources of information such as secondary data sources can also be used to examine trustworthiness of PRA data. Different PRA methods can be used to test and verify a piece of information. For example, semi-structured interviews can be supplemented by farm maps, livelihood analysis and flow charts so as to check and cross-check the information obtained. For approaching any topic the methodological design should be such that the method fits well with the problem under consideration. When it is a study of features, trends, characteristics and conditions surrounding people, quantitative method such as survey can be of considerable use. However PRA techniques which are exploratory and analytical prove to be useful when data required to be collected related to conditions which are ever changing and never stationary and are causal process involved in such conditions are dynamic.

Other Criteria Suggested

It is often said that participatory methods constitute inquiry which is undisciplined and sloppy and involve only subjective observations and are not amendable to rigor and accuracy. It is often pointed out that if information is changeable, locally valid, and value-laden; then how much of trust worthiness can the information have? There are a set of four questions which users of participatory methods attempt to answer in order to show the trustworthiness of their findings. These questions are related to the following:

- Confidence about the 'truth' of the findings;
- Applying these findings to other contexts and people;
- Replication of these findings in similar situations; and
- Certainty about the findings being determined by the subjects of inquiry rather than the biases, motivations and perspectives of the investigators.

Pretty (1993) is of the view that alternative criteria are required are required to establish trustworthiness in participatory inquiry these relate to credibility, transferability, dependability, and conformability. For example, application of participatory inquiry without triangulation of sources, techniques and investigators and participant checking of outputs will be considered as untrustworthy. The criteria listed in Pretty's (1993) paper are described below.

A framework for judging trustworthiness of data and findings can be worked out in the following ways:

- ☆ Prolonged and/or Intense Engagement Between the Various Actors in order to build trust and rapport
- ☆ Persistent and Parallel Observation for understanding a phenomenon and its context
- ☆ Triangulation applied so as to cross-check information and increase the range of different people's realities- multiple sources, methods, locations, times and investigators
- ☆ Analysis and Expression of Difference so that a wide range of different actors are involved in the analysis and their perspective and realities are represented as accurately as possible
- ☆ Negative Case analysis- revising hypothesis until one hypothesis accounts for all known cases without exception
- ☆ Peer Checking involves periodical meeting with peers not directly involved in the inquiry process
- ☆ Participant Checking to test data, interpretations and conclusions with people with whom the original information was constructed
- ☆ Reports with working Hypothesis, Contextual Descriptions and visualizations- The participatory inquiry team sets out working hypothesis. Contextual Descriptions and Visualizations– the participatory inquiry team sets out working hypotheses with detailed description of the context in which they were formulated.
- ☆ Parallel investigations and Team Communications with the same system of inquiry coming up with the same or similar findings
- ☆ Reflexive Journals- these are diaries kept by individuals on a daily basis to record a variety of information about themselves so as to help remember immediate reasons for methodological decisions and interpretations
- ☆ Inquiry Audit conducted to establish the fairness of the representations by examining the process of inquiry and the end product
- ☆ Impact of Stakeholders' Capacity to know and act so that not only should the inquiry lead to action but the rear report should prompt action on the part of the readers not directly involved.

Pretty's criteria given above are to some extent overlapping and it is possible to group them broadly into internal and external criteria as given in the following Table.

Constituents	**Evaluation**	**Criteria for trustworthiness**
Process		*Internal criteria*
The subject of inquiry From Inside	*Process/Methgods*	1. Better rapport 2. Revising Hypothesis 3. Analysis of difference 6. Observation for understanding of context 7. Triangulation
		Extern
The subject of inquiry fromOutside	*Output, Process, Methgods*	6. Participant checking 7. Peer-group checking 8. Parallel investigation 9. Inquiry audit 10. Subsequent action

Pretty's criteria, as given above, relate to process, understanding of the context, objectivity of output, multiple methods, revising learning process, recording, parallel investigations, inquiry audit and the like. All these help in reducing errors in the inquiry process.

Concluding Remarks

Conventional tests are looking at an average or 'normal' picture and any deviations from the average would either be smoothen out by further averaging or by dropping observations which are considered to be abnormal or exceptional and have major deviations from the average or expected observation. In conventional tests, while estimating parameters for causal relationship, we can apply two standard criteria, the 't' test for judging the significance of the parameters and the 'R-Square' and 'R-bar-Square' tests for judging the explanatory power of the equation with the value of 'R-Square' between '0' and '1' which when close '1' has better explanatory power. The greater are the chances of the average, the greater are the chances of the average relationship being rejected both by two standard criteria mentioned above.

Many aspects of any society's behavior can be quantified whereas other can be expressed mainly in quantitative terms. For instance, while assessing the public distribution system so as to enable the vulnerable section of society to have access to food, the general yardstick used in data indicating off take of food grains from ration shops which are retail outlets of the system. This data, does not however, capture the quality of food-grains distributed, the level of satisfaction on the poor or nutritional level of the poor. If we use qualitative method it may be widely found that the quality of food-grains supplied is not of acceptable standard such as damped wheat or broken rice or does not have the required nutritional value.

In PRA, each view is important. However, it does not obstruct the analysis in terms of an average picture. One can always look for average relationship for specific purposes. The essence is that search for an average should not tend to suppress the other relationship which may be equally important in understanding the whole gamut of relationship existing in a system. Hence, one should be aware of the missing relationship and links in the system so as to have a total understanding of the system.

Summary

In this chapter, the two systems of inquiry, questionnaire survey and PRA have been defined, illustrated and compared in terms of in terms of their paradigmatic bases, principles, objectives and ways of evaluation data. One major conclusion from this chapter is that the two systems of inquiry are based on two strikingly different 'world views' which determine their paradigms, principles and trustworthiness of their respective output. The belief system underlying questionnaire survey is that there are universal laws operating in nature which are present in samples and are to be identified and the data validated on that basis. However, in our search for an average picture or relationship we should not tend to suppress the other relationships which may be equally important in understanding the whole gamut of relationship existing in a system. It is important to appreciate such relationships from the perception of those who experience them and are an integral part of the system. In this context, we should be aware of the missing relationships and links in the system so asto attempt a total understanding of the system from different points of view. It is here that PRA being contextual and exploratory by nature evokes people's perceptions and criteria and encourages multiplicity and diversity of views leading to people's action.

Every field inquiry has of certain elements to follow, in participatory inquiry process basically being a psychological process, has been tuned to the following operating components.

The elements for field inquiry:

- ☆ Its significance
- ☆ Its nature and scope
- ☆ Its purpose
- ☆ Kinds of queries and probing
- ☆ Form/forms of output
- ☆ For whom is he output?
- ☆ Designing a plan for inquiry
- ☆ Time required
- ☆ Assessing training needs
- ☆ Methods/techniques used
- ☆ Process in field inquiry
- ☆ Cross checking of data and process
- ☆ Techniques of cross checking

- ☆ Utility of output
- ☆ Management/organization of data

Theme

To begin with any inquiry process it is important that the theme is stated and clarified whether by researchers, local community or both.

- ☆ Description of the issue/problem/theme of the field inquiry; in a participatory mode, such description if provided by local people would be more participatory, realistic and of direct relevance to the project.

Significance

Once a theme is stated, its significance also needs to be described whether by researchers or the local community involved. Much depends on whether the theme is significant at the national level or is locally significant.

- ☆ Description of the significance of the issue/problem/theme; such significance if described by local people would reflect their perceptions and views and initiate local perception from the very start. Or alternatively, to initiate a participatory-type of inquiry, the local people would need to be motivated on a theme which is not of much relevance to them.

Nature and Scope of Inquiry

- ☆ The issue and its significance would determine the nature of inquiry whether exploratory, topical, evaluative or of other kind. In PRA, the nature of inquiry can also be suggested by local people and led and conducted by them if the purpose of the field inquiry is properly clarified to them.
- ☆ For kind of output required- is it statistical representation with high degree of accuracy? Is it levels or process of change or both?/what is the unit of analysis-individuals, households, communities? And what kind of community involvement is expected in data collection and analysis.

Purpose

- ☆ On the basis of any issue and its significance as suggested by researchers/local people, the purpose of inquiry can be described by the research organization/local people, or jointly by both and its objective(s) formulated. If the objectives of the organization for the project are already existing they can be revised in the light of local people's wisdom to the extent possible.

Kind of Queries and Probing

- ✰ Description of the questions to be asked, the output required and the user-groups can be determined by researchers/local people or jointly in collaboration with the organization/researchers involved.

Forms of Output

The different stages of field inquiry involved can evoke contributions of local groups and of outside organization. The kind of output may or may not be pre-determined by researchers/local community in terms of descriptive statistics/analytical data. It can be quantitative/qualitative or a combination of both depending on the needs of the organization and the preferences and capabilities of the local people.

For Whom is the Output?

Description of the form in which the output/data is to be organized and accessed or the management information system required by the local people/ local institutions/researchers or by others. For this the relevant questions are, is the output to be taken away from the local community and processed further or is to be retained at the local level for initiating further action or is it a combination of both?

Time/Manpower/Fund Required

Description of the time required/available, manpower required/available and fund required/available for the inquiry and the contributions which local community and outside researchers can made, if necessary.

Designing a Pan for Inquiry

Definition of the area of coverage/complete enumeration/sampling frame the size of the area and the sampling procedure to be followed whether probabilistic sampling or non-probabilistic sampling. In PRA type inquiry, non-probabilistic sampling for identifying sampling units like regions, districts, blocks, villages and focus groups may be more effective to take care of the differences existing between the sampling units.

(Source and edited: Participatory Rural appraisal and questionnaire Survey by Neela Mukherjee, Concept Publishing Company, New Delhi 110059)

REFERENCES

Adhikarya, Ronny, (1994) *Strategic Extension Campaign: A Participatory Oriented Method of Agricultural Extension:* A Case Study of FAOs Experiences. Agricultural Education and Extension Services (ESHE), Human Resources, Institutions and Agrarian Reforms Division of FAO of UN, P. 209

Chambers, Roberts, (1991), "Shortcut and Participatory Methods for gaining Social Information for Projects" in Cernea Michael M. (ed.) *Putting People First, Sociological Variables in Rural Development,* Oxford University Press.

Chambers, Roberts, (1993) 'Participatory Rural Appraisal (PRA): Challenges, Potential and Paradigm', Institute of Development Studies Sussex, July.

Guba, Egon G. (ed.), (1990), *The Paradigm Dialog, Sage,* California.

Katzer, Jeffrey, Cook, Keneth H. andCrouch Wayne W. (1982). *Evaluation Information, A Guide for Users of Social Science Research,* Addisn Wesley Publishing Company, Philppines.

Pretty, N Jules (1993), "Alternative Systems of Inquiry for a Sustainable Agriculture", Sustainable Agriculture program, International Institute for Environment and Development, London 10th December, 1993.

Schwandt, Thomas R. (1990), 'Paths to Inquiry in Social Disciplines – Scientific, Constructivist and Critical Theory Methodologies", in GubaEgon G. (ed.) *The Paradigm Dialog, Sage,* California.

Streiffeler, Friedhelm, (1990), 'Qualitative Versus Quantitative Methods Method in Social Science Research', in Hoeper Bernhard (compiles) *'Qualitative Versus Quantitative Approaches in Applied Empirical Research in Rural Development,* Zed, Bonn.

— *Chapter 8* —

Reference Writing

8.1 Introduction

Sometimes writing reference list could be so confusing since there are a lot of different rules. Each universities or even each faculty use different rule in writing reference list. The different rule of writing reference list is normal because they refer to different style guide. There are a lot of styles guide available. MLA (Modern Language Association of America) style; the Chicago Manual of Style; the Oxford Guide to Style are some examples of style guide. Difference reference of style guides chosen will result in totally different way of writing reference list. The most widely accepted style guides in social science is the APA (American Psychological Association) style. In some higher education journals (*e.g.* Journal of Studies in International Education, Higher Education) it is stated clearly that author must submit their paper according to APA style. Referencing in a general sense means to give credit to someone for using his or her own ideas or thoughts in a research activity. Referencing helps in gaining the originality of the ideas and thoughts used in the research activity. Failure to reference is treated as disrespect to the original author or writer and seen as a major misconduct in the area of academic research writing. Generally students made the mistake of not mentioning proper referencing at the end of their research projects, essays or any other piece of work. This may lead to cancellation of the written matter. If we do not reference properly the written matter is treated as if it is copied from somewhere. The research paper or material cannot be submitted for further research or reading purpose. We find a lot of data when we search for something. But we do not use the data as it is. We read, study and analyze the data and then after we start preparing our thoughts about what we have to write actually. We only write relevant matter which is useful for our topic. When the exact and correct matter fits into the topic it increases the importance of what we want to say and want people to understand about the topic. So when our topic shows relevance with the matter and is acknowledged by people then who is benefitted the most. Obviously, we are the one who is benefitted from good referencing. It is essential to give reference to the author or writer of also legally advised to mention the original author name so that copyright infringement issues can be avoided.

8.2 Basic Rules of Reference List

Reference list is needed to enable readers to find the same sources used in a publication. Therefore, the data should be correct and complete. Each list contains: author name, year of publication, title and publishing data. Reference list must be double-spaced and entries should have a hanging indent. Entries should be also arranged in alphabetical order. Reference with the same author is arranged by year of publication, the earlier first. All title should be written in sentence-case (capital letter only in the first letter of first word). Only title of periodical is written in title-case (capital letter in the first letter of every word).

8.2.1 Reference List in APA Style

When writing a reference list in the APA style:

- ✰ Arrange your list alphabetically by author's surname.
- ✰ Use and before the last author if there are 2 – 7 authors.
- ✰ When there are 8 or more authors, write the first 6 followed by an ellipsis and add the last author. Example: Krishnan, K. J., Reeve, A. K., Samuels, D. C., Chinnery, P. F., Blackwood, J. K., Taylor, R. W.,. Turnbull, D. M.
- ✰ Add an English translation of the title in square brackets if the source used is not in English.
- ✰ Use italics for:
  - ❒ Titles of journals and volumes
  - ❒ Book titles
- ✰ Indent the paragraph on the second and the following lines in a reference.
- ✰ Start the reference list on a new page. Use "Reference list" or "Literature list" as the heading.

8.2.2 The APA Style in-text

When using the APA style in-text:

- Multiple publications by the same author published the same year are distinguished by a, b, c *etc.* after the year: Hansen (1988a) and Hansen (1988b)
- When a work has no identifiable author, use the first words of the title and year of publication. Article titles and chapters are written with quotation marks, but book titles, pamphlets or reports are written in italics. Example: *College Bound Seniors* (2008)
- Publications with multiple authors to which you refer several times in the text:
  - 2 authors: name both authors each time you quote them. Example: Furseth and Everett (1997) – or – (Furseth and Everett, 1997)
  - 3, 4 or 5 authors: name all authors the first time you quote them. From then on use only the surname of the first author, followed by *et al.*, Example: Ramaekers, Berghausb, Laarc and Drummer (2004), followed by Ramaekers *et al.* (2004)
  - 6 or more authors: only give the first author's surname, followed by *et al.*, Example: Cheng *et al.* (2004)
- When using secondary references, name your source and cite the secondary reference. Example: Johnson and Peters' studies (as cited in Wagner, 1982)...
- Give page numbers:
  - always for direct quotations
  - page numbers are encouraged for indirect quotations or for citing a specific part of a source, especially when it would help the reader locate the relevant passage in a long text, for instance a book

8.3 Importance of Referencing

Making proper references provides various benefits. Some of its advantages can be cited as follows:

The major benefit of good referencing is that it authenticates the originality of the source from where we have taken our subject matter. The matter can be in the form of words, visuals, audios, numerals, facts *etc.* It also shows how deeply we have understood the topic on which we are working. It also recognizes the writer and his work because he or she should also be accredited for what they have done. It is an act to pay respect for the hard work and dedication and the efforts one has put into the research work.

When we provide references our readers can also analyze the information given by us on the basis of the original source. They can cross check the details from the original document. This helps us them to make an understanding of what we are trying to say in the research project and how far we are able to achieve our goal. It also helps us to support our facts in favor or in against of the said topic. When we present the true facts people can judge that how correctly we are to comprehend things in a dynamic way without losing the originality of the topic under consideration.

Sometimes readers want to go beyond what we have mentioned in our research projects. They want to go further in the topic and want some more information. Referencing helps a lot in this as researchers can be provided with more information of their use from the original source.

Referencing is also useful as it avoids the chances of plagiarism which is a type of copyright infringement issue. Using somebody else's data without his or her permission and putting our own name for the work done is known as plagiarism. Most universities and colleges cancels students project because of plagiarism. Plagiarism is considered as a legal offence. Universities and colleges have an anti plagiarism software so that it can be detected which students have plagiarism related issue in their project.

Referencing allows you to acknowledge the contribution of other writers and researchers in your work. Any university assignments that draw on the ideas, words or research of other writers must contain citations.

Referencing is also a way to give credit to the writers from whom you have borrowed words and ideas. By citing the work of a particular scholar, you acknowledge and respect the intellectual property rights of that researcher. As a student or academic, you can draw on any of the millions of ideas, insights and arguments published by other writers, many of whom have spent years researching and writing. All you need to do is acknowledge their contribution to your assignment.

Referencing is a way to provide evidence to support the assertions and claims in your own assignments. By citing experts in your field, you are showing your marker that you are aware of the field in which you are operating. Your citations map the space of your discipline and allow you to navigate your way through your chosen field of study, in the same way that sailors steer by the stars.

References should always be accurate, allowing your readers to trace the sources of information you have used. The best way to make sure you reference accurately is to keep a record of all the sources you used when reading and researching for an assignment.

8.4 Various Types of Referencing

Referencing itself includes a number of different types of styles which are as follows:

☆ APA (American Psychological Association)

The APA style of referencing is generally known as author date style. In this style author name is mentioned along with date and page number. This style is used at Polytechnic West. Two basic requirements of APA style are in-text citations and end-text citations. In-text citations is shown in the following format-

(Author's name, year or date of publication)

It includes information from all the resources which we have used within our document like journals, magazines, research paper *etc.* This is generally used when we quote some comment or quotation in our assignment. For example, Shakespeare said, "This world is a stage and we all are playing our roles here." This can be said like this,

This world is a stage and we all are playing our roles here. (Shakespeare, year in which the official statement was written 1942)

☆ MLA

In MLA style of referencing only name of the author along with page number is shown. Here, year of referencing is not shown as in APA style of referencing. For example, (Shakespeare 56).Also, commas are not used in this.

☆ Harvard

Harvard style of referencing is very popular among college and university students. In this also the two basic requirements is in-text citations and reference list. When we quote something directly from a resource it is in-text citations. When we mention the resources referred in our research at the end of our assignment it is known as reference list. It also gives the full description of the sources which we have used in our project.

☆ Chicago

The Chicago style of referencing does not use in text citations and reference lists. It uses notes and bibliography and also author date. Bibliography is used when we mention each and everything that we have used in our research report. It gives detailed information about what resources we have used in our complete research project. That's why it is mentioned at the end of the report.

APA (American Psychological Association) style is used in the social sciences, arts and humanities. Check which reference style your department recommends before you begin writing your paper.

8.5 Reference List

A reference list lists only the sources you refer to in your writing. The purpose of the reference list is to allow your sources to be found by your reader. It also gives credit to authors you have consulted for their ideas. All references cited in the text must appear in the reference list, except for personal communications (such as conversations or emails) which cannot be retrieved.

A bibliography is different from a reference list as it lists all the sources used during your research and background reading, not just the ones you refer to in your writing.

8.5.1 Reference Formatting Guide

Title	Include the title 'References' (one word, beginning with a capital letter, centred, and not in italics
Indent	Hanging indent your references (space bar in 5 - 7 spaces for the second and subsequent lines of each reference)
Space between references	In general double-space between references
Ampersand	Use for 2 - 6 authors, use " and " before the final author
One author, two publications	Order by year of publication, the earlier one first. Same year of publication for both - add 'a' and 'b' after the year, inside the brackets. Include this in the in text citation. example: Baheti, J. R. (2001a).
URLs	Remove the underlines from URLs so that any underscores (_) can be seen
Same first author, different second author	Order alphabetically by second or subsequent authors
Upper case letters (capital letters)	Journal title - use headline style; *i.e.* capitalise all the words, except articles and prepositions Book title or article title (in a journal, magazine or newspaper) - use sentence style; *i.e.* capitalise the first word of the title, and subtitle (after the colon), and any proper names
Place of publication	USA publishers give the city in full and the abbreviation for the state. New York, NY Springfield, MA Publishers outside the USA: Give the city in full and the country in full London, England Auckland, New Zealand
Page range	Use an en dash, NOT a hyphen, for page ranges: *e.g.* 21–27. No gaps between the page numbers and the en dash. How to add an en dash in Microsoft Word, if using a full PC keyboard: Hold the ***Control key*** and type the ***minus sign*** on the small numeric keypad. **NB:** If your keyboard will not produce an en dash, it is acceptable to use a hyphen instead. See the *Publication manual of the American Psychological Association* (2010, p. 97) for more detail on the use of hyphens and dashes in APA style.

Use of square brackets	If format, medium or description information is important for a resource to be retrieved or identified, use square brackets after the title to include this detail: Scorsese, M. (Producer), and Lonergan, K. (Writer/Director). (2000). *You can count on me* [Motion picture]. United States: Paramount Pictures.

8.5.2 Reference Examples

One author (a book chapter)

Easton, B. (2008). Does poverty affect health? In K. Dew and A. Matheson (Eds.), Understanding health inequalities in Aotearoa New Zealand (pp. 97–106). Dunedin, New Zealand: Otago University Press.

One author, multiple works published in the same year

Rush, E., McLennan, S., Obolonkin, V., Cooper, R., and Hamlin, M. (2015a). Beyond the randomised controlled trial and BMI—evaluation of effectiveness of through-school nutrition and physical activity programmes. *Public Health Nutrition, 18*(9), 1578–1581. https://doi.org/10.1017/S1368980014003322

Rush, E. C., Obolonkin, V., Battin, M., Wouldes, T., and Rowan, J. (2015b). Body composition in offspring of New Zealand women: Ethnic and gender differences at age 1–3 years in 2005–2009. *Annals Of Human Biology, 42*(5), 492–497.

Two authors (a journal article with doi)

Li, S., and Seale, C. (2007). Learning to do qualitative data analysis: An observational study of doctoral work. *Qualitative Health Research, 17*(10), 1442-1452. https://doi.org/10.1177/1049732307306924

Three authors

Barnard, R., de Luca, R., and Li, J. (2015). First-year undergraduate students' perceptions of lecturer and peer feedback: A New Zealand action research project. *Studies In Higher Education, 40*(5), 933–944. https://doi.org/10.1080/03075079.2014.881343

Use " and " before the final author.

Four to seven authors

Szcz Ê Sna, A., Nowak, A., Grabiec, P., Paszkuta, M., Tajstra, M., and Wojciechowska, M. (2017). Survey of wearable multi-modal vital parameters measurement systems. *Advances in Intelligent Systems and Computing, 526*. https://doi.org/10.1007/978-3-319-47154-9_37

List all authors in the reference entry

More than seven authors

Kasabov, N., Scott, N. M., Tu, E., Marks, S., Sengupta, N., Capecci, E.,. Yang, J. (2016). Evolving spatio-temporal data machines based on the NeuCube neuromorphic framework: Design methodology and selected applications. *Neural Networks, 78,* 1-14. https://doi.org/10.1016/j.neunet.2015.09.011

First 6 authors. last author. and follow by date and other information.

Secondary citations

A secondary citation is where you are citing information or quotes the author of your reference has taken from source that you have not read.

In-text citation

Seidenberg and McClelland's study, conducted in 1990 (as cited in Coltheart, Curtis, Atkins, and Haller, 1993), shows that.

. as some studies show (Seidenberg and McClelland, as cited in Coltheart, Curtis, Atkins, and Haller, 1993).

Reference list entry

Coltheart, M., Curtis, B. Atkins, P., and Haller, M. (1993). Models of reading aloud: Dual-route and parallel-distributed-processing approaches. *Psychological Review,* 100, 589–608.

DOI

A DOI, or digital object identifier, is a unique, permanent identification number that will take you straight to a document no matter where it is located on the Internet. Note that a DOI will usually link to a record on a publisher's website, and may not always include full text, even though the Library may have full text access. It's always worthwhile checking the Library catalogue or databases and e-journals page to see if full text is available.

DOI = digital object identifier

- ☆ A DOI commonly identifies a journal article but it can also be found on other publication types including books.
- ☆ All DOIs start with 10. and includes numbers and letters. Example: doi:10.1016/j.addbeh.2009.08.001
- ☆ The DOI provides a permanent internet address for the item making it easy to locate.
- ☆ You may search by DOI numbers in Library Search to locate articles.

Doi in your reference list entry

- ☆ **Always** use the DOI if available (for print or online articles and books).
- ☆ **No** full stop at the end of a DOI.

New!

A new citing format for DOI was introduced by APA in March 2017. The new format includes *https* and the prefix *doi.org*: **https://doi.org/10.1016/j.asw.2016.11.001**

Example

Oppenheimer, D., Zaromb, F., Pomerantz, J. R., Williams, J. C., and Park, Y. S. (2017). Improvement of writing skills during college: A multi-year cross-sectional and longitudinal study of undergraduate writing performance. *Assessing Writing*, 32, 12–27. https://doi.org/10.1016/j.asw.2016.11.001

NOTES

Articles retrieved from library databases may include **ezproxy.aut.ac.nz** in the DOIs. This ezproxy information should be removed.

For example:

https://doi-org.~~ezproxy.aut.ac.nz~~/10.1093/pubmed/fdv045

The correct URL for this DOI is:

https://doi.org/10.1093/pubmed/fdv045

URL

If there is no DOI for a online journal article or an e-book, include a URL in your reference.

Use the URL of a journal home page for journal articles without DOI

- ☆ Use the URL of the journal homepage, NOT the full URL of the article, in your reference.

Finding a journal homepage URL:

- ☆ You could do a Google search for the journal title (within double quotation marks), *e.g.* "new zealand management magazine" to find the journal's homepage
- ☆ Or, go to the Library database Ulrichsweb, search by the journal title or the journal's ISSN to find the journal record. On the journal record page, find the journal URL for your reference.

Journals without a home page and no DOI

This can happen to some discontinued journals, or journals archived in an archival database only.

- ☆ Use the **database home page** URL in your reference. See the example in the following section.

Use a URL of a library database

Resources retrieved from a library database, without a DOI:

If you use electronic resources without DOI, such as an ebook or a data set or a journal without a website, from a library database, You are required to include the URL of the *database homepage* in your reference.

- ☆ Do not use the full URL of the source that you retrieved from a database.

Example

An e-book *"Small town sustainability: economic, social, and environmental innovation"*.

The URL on the e-book page is:

https://ebookcentral-proquest-com.ezproxy.aut.ac.nz/lib/AUT/detail.action?docID=1121624

- ☆ You should remove ezproxy and other details following it.
- ☆ The correct URL in the reference for this book should be http://ebookcentral.proquest.com

Quotations

Direct quotations shorter than 40 words are integrated in the text and placed within quotation marks. Quotations of 40 words or more should be in a separate indented paragraph, without quotation marks.

Examples

- ☆ Short quotation: "It is permissible to quote, word for word, from a source, but in most disciplines, this should be done sparingly" (Day, 2013, p. 143).
- ☆ The name of the author is integrated in the paragraph: Day (2013) claims that "it is permissible to quote, word for word, from a source, but in most disciplines, this should be done sparingly" (p. 143).
- ☆ Source with more than one author: "The purpose of referencing is related to the ideal of research as a collective project" (Furseth and Everett, 2013, p. 139).

Indirect quotations - paraphrases

An indirect quotation (paraphrase) is a reformulation of the original text.

Example: Furseth and Everett (1997) maintain that the primary reason behind the use of references and bibliography is the idea of research as a collective endeavour. Research should be verifiable, and those reading your work should be able to find those sources your material is based upon (p. 142).

Personal Communications

Personal communications may be telephone conversations, e-mails, and the like. As they do not provide recoverable data, do not include personal communications in the reference list. Cite personal communications in text only, and get permission of the person concerned. Example: (T. K. Smith, personal communication, April 18, 2016)

8.6 Structure/Template of Different Article

1. Structure/Template of BOOK

Surname, Initial(s). (Year). *Book title in italics* [Translated title – if applicable] (Edition - if available). Place: Publisher.

Add an English translation of the title in square brackets if the source used is not in English.

In a reference list

Creme, P., and Lea, M. R. (2003). *Writing at university* (2nd ed.). Maidenhead: Open University Press.

In text

(Creme and Lea, 2003, p. 10)

2. Structure/Template of E-book

Surname, Initial(s). (Year). *Book title in italics* [Translated title – if applicable]. DOI

Surname, Initial(s). (Year). *Book title in italics* [Translated title – if applicable]. Place: Publisher. Retrieved from https://.

Add an English translation of the title in square brackets if the source used is not in English.

In a reference list

Wilbur, J. (2014). *A grammar of Pite Saami*. https://doi.org/10.17169/langsci.b17.34

In text

(Wilbur, 2014)

3. Structure/Template of Book Chapter

Surname author, Initial(s). (Year). Title of chapter [Translated title – if applicable]. In Initial(s) Surname editor (Ed.), *Book title in italics* (Edition - if available, page). Place: Publisher.

Add an English translation of the title in square brackets if the source used is not in English.

In a reference list

Beizer, J. L., and Timiras, M. L. (1994). Pharmacology and drug management in the elderly. In P. S. Timiras (Ed.), *Physiological basis of aging and geriatrics* (2nd ed., pp. 279-284). Boca Raton, FL: CRC Press.

In text

(Beizer and Timiras, 1994)

4. Structure/Template of Book with an Editor

Surname editor, Initial(s). (Ed.). (Year). *Book title in italics* [Translated title – if applicable] (Edition – if available). Place: Publisher.

Add an English translation of the title in square brackets if the source used is not in English.

In a reference list

Fermann, G. (Ed.). (1997). *International politics of climate change: key issues and critical actors.* Oslo: Scandinavian University Press.

In text

(Fermann, 1997)

5. Structure/Template of Journal Article

Surname, Initial(s). (Year). Article title [Translated title – if applicable]. *Title of journal in italics, volume in italics*(number), page. DOI – if available

Surname, Initial(s). (Year). Article title [Translated title – if applicable]. *Title of journal in italics, volume in italics*(number), page. Retrieved from https://.

If retrieved from the Internet and no DOI has been assigned, give URL. Give the URL of the journal's web site.

Add an English translation of the title in square brackets if the source used is not in English.

In a reference list

Kwan, I., and Mapstone, J. (2004). Visibility aids for pedestrians and cyclists: A systematic review of randomised controlled trials. *Accident Analysis and Prevention, 36*, 305-312. https://doi.org/10.1016/S0001-4575(03)00008-3

Issue number is not included as the journal uses continuous pagination.

Kaslegard, B. (2018). Kvar er «heime»? [Where is "home"?]. *Samtiden, 128*(1), 72-77. Retrieved from https://samtiden.no

In text

(Kwan and Mapstone, 2004)

(Kaslegard, 2018)

6. Structure/template of Conference Paper

Surname author, Initial(s). (Publication year). Title of paper [Translated title – if applicable]. In Initial(s) Surname editor (Ed.), *Conference Name in italics* (page). Place: Publisher.

Add an English translation of the title in square brackets if the source used is not in English.

In a reference list

Nørvåg, K. (2003). Space-efficient support for temporal text indexing in a document archive context. In T. Koch and I. Sølvberg (Eds.), *Research and Advanced Technology for Digital Libraries, 7th European Conference, ECDL 2003 Trondheim, Norway* (pp. 511-522). Berlin: Springer.

In text

(Nørvåg, 2003)

7. Structure/Template of Newspaper Article

Surname, Initial(s). (Year, Date). Title of article [Translated title – if applicable]. *Title of newspaper in italics*, page.

Surname, Initial(s). (Year, Date). Title of article [Translated title – if applicable]. *Title of newspaper in italics*. Retrieved from https://.

To reference an online newspaper article, give URL. Give the URL of the newspaper's web site when the article is available by search, *e.g.* Retrieved from https://www.aftenposten.no

Add an English translation of the title in square brackets if the source used is not in English.

In a reference list

Brody, J. E. (2007, December 11). Mental reserves keep brain agile. *The New York Times*. Retrieved from https://www.nytimes.com

In text

(Brody, 2007)

8. Structure/Template of Dissertation/thesis

Surname, Initial(s). (Year). *Title of dissertation or thesis in italics* [Translated title – if applicable] (Doctoral dissertation). Institution, Place.

Surname, Initial(s). (Year). *Title of dissertation or thesis in italics* [Translated title – if applicable] (Doctoral dissertation, Institution). Retrieved from https://.

Add an English translation of the title in square brackets if the source used is not in English.

In a reference list

Hasund, I. K. (2003). *The discourse markers like in English and liksom in Norwegian teenage language: A corpus-based, cross-linguistic study* (Doctoral dissertation). University of Bergen, Bergen.

In text

(Hasund, 2003)

9. Structure/Template of Report

Surname, Initial(s). (Year). *Title of the report in italics* [Translated title – if applicable] (Report No.). Place: Publisher.

Surname, Initial(s). (Year). *Title of the report in italics* [Translated title – if applicable] (Report No.). Retrieved from https://.

To reference a report retrieved online, replace information about place and publisher by URL.

Add an English translation of the title in square brackets if the source used is not in English.

In a reference list

Amundsen, I. (2010). *Support for parliaments: Tanzania and beyond* (CMI report R 2010:8). Retrieved from http://hdl.handle.net/11250/2436092

In text

(Amundsen, 2010)

10. Structure/Template of Web page

Surname, Initial(s). (Year). Title [Translated title – if applicable]. Retrieved from https://.

Give publication date or use the date on which the content was last updated.

Do not include retrieval date unless the web page content is subject to frequent change. If needed, use this format: Retrieved October 20, 2017, from https://.

Add an English translation of the title in square brackets if the source used is not in English.

In a reference list

Nielsen, J. (2012, January 4). Usability 101: Introduction to usability. Retrieved from https://www.nngroup.com

In text

(Nielsen, 2012)

11. Structure/Template of encyclopedia

With an identified author:

Surname, Initial(s). (Year). Title of the entry [Translated title – if applicable]. In Initial(s) Surname editor (Ed.), *Title of the encyclopedia in italics* (Edition - if available, page). Retrieved from https://.

Add an English translation of the title in square brackets if the source used is not in English.

Without an identified author:

Title of the entry. (Year). In Initial(s) Surname editor (Ed.), *Title of the encyclopedia in italics* (Edition - if available, page). Retrieved from https://.

In a reference list

Without an identified author:

Psychology. (n.d.). In *Wikipedia*. Retrieved October 14, 2009, from http://en.wikipedia.org/wiki/Psychology

Retrieval date is included in this reference because the content may change over time.

As no publication date is available, n.d. (for "no date") is entered.

In text

Without an identified author:

("Psychology," n.d.)

12. Structure/Template of Blog Post

Surname, Initial(s). (Year, Date published). Title of the blog post [Translated title – if applicable] [Blog post]. Retrieved from http://.

In a reference list

Mathis, T. (2015, August 12). What is human systems integration? [Blog post]. Retrieved from http://blog.apabooks.org/2015/08/12/what-is-human-systems-integration

In text

(Mathis, 2015)

13. Structure/Template of Image

Surname artist, Initial(s). (Year). *Title of work in italics* [Translated title – if applicable] [Medium]. Place: Institution.

Surname artist, Initial(s). (Year). *Title of work in italics* [Translated title – if applicable] [Medium]. Retrieved from http://.

If the institution has a web site where you can view the painting, give the URL of the web site, *e.g.* Retrieved from https://www.moma.org/collection/

In a reference list

Condo, G. (2009). *The fallen butler* [Painting]. New York: Museum of Modern Art.

In text

(Condo, 2009)

8.7. What is an Abstract?

An abstract is a short summary of a larger work, such as a dissertation or research paper. It allows potential readers to quickly identify what your paper is about and decide if it's worth reading. All abstracts should include:

- ☆ A statement of your main topic, purpose and objectives
- ☆ A brief description of the methodology
- ☆ An overview of the most significant findings or arguments
- ☆ A summary of your conclusions and recommendations

The abstract appears at the very beginning of a document, but it should be the last thing you write. In a dissertation or thesis, include it on a separate page, after the title page and acknowledgements but before the table of contents.

Most abstracts are around 150-300 words, but the length depends on the requirements of your assignment—often you will be given a strict word limit.

8.7.1 Tips for Writing an Abstract

It can be a real challenge to condense your whole dissertation into just a couple of hundred words, but the abstract will be the first (and sometimes only) part that people read, so it's important to get it right. These strategies can help you get started.

Reverse Outline

Not all abstracts will contain precisely the same elements. If your research has a different structure (for example, a humanities dissertation that builds an argument through thematic chapters), you can write your abstract through a process of reverse outlining.

For each chapter or section, list keywords and draft 1-2 sentences that summarize the central point or argument. This will give you a framework of your abstract's structure. Next, revise the sentences to make connections and show how the argument develops.

The abstract should tell a condensed version of the whole story, and it should only include information that can be found in the main text. Reread your abstract to make sure it gives a clear summary of your overall argument.

Read other Abstracts

The best way to learn the conventions of writing an abstract is to read other people's. You probably already read lots of journal article abstracts while conducting your literature review—try using them as a framework for structure and style.

Write Clearly and Concisely

A good abstract is short but impactful, so make sure every word counts. Each sentence should clearly communicate one main point.

Avoid unnecessary filler words, and avoid obscure jargon that requires explanation—the abstract should be understandable to readers who are not familiar with your topic.

Focus on your own Research

The purpose of the abstract is to report the original contributions of your research, so avoid discussion of others' work, even if you address it at length in the main text.

You might include a sentence or two summarizing the scholarly background to situate your research and show its relevance to a broader debate, but there's no need to mention specific publications. Don't include citations in an abstract unless absolutely necessary (for example, if your research responds directly to another study or revolves around one key theorist).

Check your Formatting

If you are writing a thesis or dissertation or submitting to a journal, there are often specific formatting requirements for the abstract—make sure to check the guidelines and format your work correctly. For APA research papers you can follow the APA abstract format.

Always stick to the word limit. If you have not been given any guidelines on the length of the abstract, it's best not to write more than one page.

8.7.2 Checklist: Abstract

- ☆ The word count is within the required length, or a maximum of one A4 sheet of paper.
- ☆ The abstract is included after the title page and/or acknowledgements and before the table of contents.
- ☆ The objective, problem statement, and/or research questions are stated.
- ☆ The methodology is briefly described.
- ☆ The most important results are summarized.
- ☆ The main conclusions are stated.
- ☆ Any recommendations and/or suggestions for further research are mentioned.
- ☆ The abstract can be understood by someone without prior knowledge of the topic.

8.8 Plagiarism

8.8.1 What is Plagiarism?

Plagiarism at UNSW is using the words or ideas of others and passing them off as your own. Plagiarism is a type of intellectual theft.

Plagiarism can take many forms, from deliberate cheating to accidentally copying from a source without acknowledgement. Consequently, whenever you use the words or ideas of another person in your work, you must acknowledge where they came from.

8.8.2 Three Steps to Learning about Plagiarism

It's important to know what plagiarism is, and what form it takes (some common types of plagiarism are listed here). It's also important to **know how plagiarism happens.** The final step is to **develop effective academic skills.** Many students who plagiarise do so unintentionally, often because they don't have the academic skills to avoid over-reliance on the work of others or because they aren't sure what constitutes plagiarism. So it's important to take every opportunity to **develop academic skills.** On this site, there are resources on avoiding plagiarism and how to be organized, as well as a list of other resources and links so students can develop good academic practice.

8.8.3 Why is it Wrong to Plagiarise?

Plagiarism is unethical for three reasons:

- ☆ Firstly, it is unethical because it is a form of theft. By taking the ideas and words of others and pretending they are your own, you are stealing someone else's intellectual property.

- Secondly, it is unethical because the plagiariser subsequently benefits from this theft.
- Thirdly, a degree is evidence of its holder's abilities and knowledge. If a student gains employment on the basis of a qualification they have not earned, they may be a risk to others.

No doubt some students do cheat. They deliberately take the results of other people's hard work, use it to gain credit for themselves, and learn little or nothing in the process. But most cases of plagiarism are accidental and could be avoided if students became more conscious of their own writing and research practices. Most students who plagiarise do so unintentionally, usually because they don't have the skills to avoid over-reliance on the work of others or because they aren't sure what constitutes plagiarism.

8.8.4 What is h-index?

The Hirsch index (h-index) is an author-level metric that attempts to measure both the productivity and citation impact of the publications of a scientist or a scholar. The **index** is based on the set of the scientist's most cited papers and the number of citations that they have received in other publications.

Formally, if f is the function that corresponds to the number of citations for each publication, we compute the h-index as follows. First we order the values of f from the largest to the lowest value. Then, we look for the last position in which f is greater than or equal to the position (we call h this position). For example, if we have a researcher with 5 publications A, B, C, D, and E with 10, 8, 5, 4, and 3 citations, respectively, the h-index is equal to 4 because the 4th publication has 4 citations and the 5th has only 3. In contrast, if the same publications have 25, 8, 5, 3, and 3 citations, then the index is 3 because the fourth paper has only 3 citations.

$f(A)=10, f(B)=8, f(C)=5, f(D)=4, f(E)=30 \rightarrow h\text{-index}=4$

$f(A)=25, f(B)=8, f(C)=5, f(D)=3, f(E)=30 \rightarrow h\text{-index}=3$

If we have the function f ordered in decreasing order from the largest value to the lowest one, we can compute the h-index as follows:

$$\text{h-index (f)} = \max_i \min(f(i), i)$$

The Hirsch index is analogous to the Eddington number, an earlier metric used for evaluating cyclists. The h-index serves as an alternative to more traditional journal impact factor metrics in the evaluation of the impact of the work of a particular researcher. Because only the most highly cited articles contribute to the h-index, its determination is a simpler process. Hirsch has demonstrated that h has high predictive value for whether a scientist has won honors like National Academy membership or the Nobel Prize. The h-index grows as citations accumulate and thus it depends on the "academic age" of a researcher.

8.8.5 What is i10-index?

The i10-index is the newest in the line of journal metrics and was introduced by Google Scholar in 2011. It is a simple and straightforward indexing measure found by tallying a journal's total number of published papers with at least 10 citations (Google Scholar Blog, 2011).

8.8.6 The Journal Impact Factor

The impact factor (IF) of a scientific journal is a measure reflecting the average number of citations to papers published in that journal. This indicator measures the relative importance of a journal within its scientific field, with journals with higher impact factors deemed to be more important than those with lower ones. The impact factor was devised by Eugene Garfield, the founder of the Institute for Scientific Information (ISI) in Philadelphia (now a part of Thomson Reuters), as a way to count the impact of scientific journals (Garfield 1964, 1972, 2006). Thomson Reuters calculates the impact factor of journals every year. On the other hand, The Scopus Journal Metrics allow for direct comparison of journals, independent of their subject classification (Elsevier B.V., 2016).

REFERENCES

Alred, G. J., Brusaw, C. T., and Oliu, W. E. (2009). *The business writer's handbook.* New York, NY: St Martin's Press.

Best, A. (2004). *International history of the twentieth century.* Retrieved from http://www.netlibrary.com

Easton, B. (2008). Does poverty affect health? In K. Dew and A. Matheson (Eds.), *Understanding health inequalities in Aotearoa New Zealand* (pp. 97-106). Dunedin, New Zealand: Otago University Press.

Elsevier B.V. (2016). Scopus Journal Metrics - Webology. Retrieved June 20, 2016, from http://www.scopus.com/source/sourceInfo.uri?sourceID=4400151723 and origin=resultlist, www.journalmetrics.com[5] Flesch, R. (n.d.). How to write plain English. Retrieved April 12, 2009, from http://www.mang.canterbury.ac.nz/writing_guide/writing/flesch.shtml.

Global warming. (2009, June 1). Retrieved June 4, 2009, from http://en.wikipedia.org/wiki/Global_warming[7] Google Scholar Blog (2011). Google Scholar citations open to all. Published 16 November 2011. Retrieved June 20, 2016, from http://googlescholar.blogspot.com/2011/11/google-scholar-citations-open-to-all.html

Li, S., and Seale, C. (2007). Learning to do qualitative data analysis: An observational study of doctoral work. *Qualitative Health Research, 17,* 1442–1452. https://doi.org/10.1177/1049732307306924

Radio New Zealand. (2008). *Annual report 2007-2008*. Retrieved from http://static.radionz.net.nz/assets/pdf_file/0010/179676/Radio_NZ_Annual_Report_2008.pdf

Read, E. (2007, November 1). Myth-busting gen Y. *New Zealand Management*. Retrieved from http://www.management.co.nz[11] Thomson Reuters (1994). The Thomson Reuters Impact Factor. Retrieved June 10, 2016, from http://wokinfo.com/essays/impact-factor/

— Chapter 9 —

Statistical Software for Data Analysis

Statistics as a wide subject is used in almost all disciplines especially in research studies. Statistics provides scientific tools for representative data collection, appropriate data analysis and summarization of data and inferential procedure for drawing valid conclusion in the face of uncertainty. Each and every researcher should have some knowledge in Statistics and must use statistical tools in his or her research, one should know about the importance of statistical tools and how to use them in their research or survey. The quality assurance of the work must be dealt with: the statistical operations necessary to control and verify the analytical procedures as well as the resulting data making mistakes in analytical work is unavoidable. The emergence of statistical software in the twenty-first century has helped different researchers in the physical and social science to improve in the quality of research. Most renowned researchers in adopting this software in their data analysis have been able to identify the immense contribution to research findings (Adetola, 2013). Any quantitative research cannot be done effectively without Statistical Software. Moreover, it enables research data for easy presentation. It helps professionals to interact with data thereby paving way for creativity and innovation. Some are user friendly interface with drop-down tips for beginners (ATS, Ucla Edu, 2014). Advances in technology have improved all our lives (Akindutire, 2013) and has allowed experts greater freedom to come out with results within a twinkle of an eye than ever before where it takes time to finish analysis. This same technology has offer tremendous opportunity to research and keep research as a more interesting field of study. This study is significant, because it is an attempt to measure the direct impact of different statistical software on research analysis.

9.1 Common Statistical Software and their Application

Stephanie D. (2009), Standford PhD Statistical Consulting and Karen (2013), Wikipedia library and other scholars have identified different popular statistical software programs, which are SPSS, Eview, SAS, MATLAB, MINITAB, STATA, Mathematica and lots more have been utilized by people across all disciplines for many years and are quite user friendly.

a) Statistical Package for the Social Sciences (SPSS)

(Statistical Package for the Social Sciences now Statistical products and Solution services) is most widely used in social science disciplines and courses. SPSS is the oldest software programs developed and made available in 1960s and has been redeveloped over the years, the latest version is SPSS 24.0. Many sociologists, psychologists and social workers use this program to enter their research data and formulate results. Although social science uses SPSS more widely than other fields, many find it easy to navigate with SPSS because it is a package that many beginners enjoy due to its very easy to use nature. SPSS has a "point and click" interface that allows you to use pull down menus to select commands that you wish to perform. Odusina (2011) disclosed that working with SPSS demand some background knowledge of statistics. There are slight variations in the difference version of SPSS *e.g.* version 10, 11, 12, 13, 14, 15, 16, 18, *etc.* SPSS assists the user in describing data, testing hypotheses and looking for a correlation or relationship between one or more variables. SPSS is very suitable for most regression analysis and different kinds of ANOVA (regression, logistic regression, survival analysis, analysis of variance, factor analysis, multivariate analysis but not suitable for time series analysis and multilevel regression analysis)-Wikipedia (2014). Many students, both undergraduate and graduate, are taught SPSS during research analysis classes in demography, psychology, sociology and other social sciences.

b) Statistical Analysis System (SAS)

SAS is a software suite developed by SAS Institute for advanced analytics, multivariate analyses, business intelligence, data management, and predictive analytics. In order to use Statistical Analysis System, Data should be in a spreadsheet table format or SAS format. SAS programs have a DATA step, which retrieves and manipulates data, usually creating a SAS data set, and a PROC step, which analyses the data. Each step consists of a series of statements. The DATA step has executable statements that result in the software taking an action, and declarative statements that provide instructions to read a data set or alter the data's appearance. The DATA step has two phases, compilation and execution. In the compilation phase, declarative statements are processed and syntax errors are identified. Afterwards, the execution phase processes each executable statement sequentially. Data sets are organized into tables with rows called "observations" and columns called "variables". Additionally, each piece of data has a descriptor and a value. The PROC step consists of PROC statements that call upon named procedures. Procedures perform analysis and reporting on data sets to produce statistics, analyses and graphics. There are more than 300 procedures and each one contains a substantial body of programming and statistical work. PROC statements can also display results, sort data or perform other operations. SAS Macros are pieces of code or variables that are coded once and referenced to perform repetitive tasks. SAS data can be published in HTML, PDF, Excel and other formats using the Output Delivery System, which was first introduced in 2007. The SAS Enterprise Guide is SAS' point-and-click interface. It generates code to manipulate data or perform analysis automatically and does not require SAS programming experience to use.SAS is one of the packages that are difficult to learn. To use SAS, one must write SAS programs that manipulate the data and perform data analyses. If you make a mistake in a SAS program, it can be hard to see where the errors occurred or how to correct it. However, it can take a long time to learn and understand data management in SAS than many other packages with simpler commands line. However, SAS can work with many data files at once SAS can handle enormous data files and the number of records is generally limited to the size of user's hard disk. SAS performs most general statistical analyses (regression, logistic regression, survival analysis, analysis of variance, factor analysis, multivariate analysis). The greatest strengths of SAS are probably in its ANOVA, mixed model analysis and multivariate analysis, while it is probably weakest in ordinal and multinomial logistic regression (because these commands are especially difficult), and robust methods (it is difficult to perform robust regression, or other kinds of robust methods)- ATS Ucla Edu(2014). While there are some supports for the analysis of survey data, they are quite limited as compared to STATA.

c) Econometric Views (EViews)

Eviews is a statistical package for Windows, used mainly for time-series oriented econometric analysis. It is developed by Quantitative Micro Software (QMS), now a part of IHS. Version 1.0 was released in March 1994, and replaced MicroTSP. The TSP software and programming language had been originally developed by Robert Hall in 1965. The current version of EViews is 9.5, released in March 2016. EViews can be used for general statistical analysis and econometric analyses, such as cross-section and panel data analysis and time series estimation and forecasting. EViews relies heavily on a proprietary and undocumented file format for data storage. EViews combines spreadsheet and relational database technology with the traditional tasks found in statistical software, and uses a Windows GUI. This is combined with a programming language which displays limited object orientation. However, for input and output it supports numerous formats, including databank format, MS-Excel format, SPSS/PSPP, DAP/SAS, STATA, RATS, and TSP. EViews can access ODBC databases.

d) MINITAB

It is statistical software used by educators, students, scientists, business associates and researchers to provide statistical software in a multitude of areas. MINITAB, developed around 1990, and remains one of the oldest statistical software programs available. MINITAB has compatibility with PC, Macintosh, Linux and all other major platforms. As one of the easiest statistical software programs to use, MINITAB remains a popular choice with those new statistical software. With drop-down menus and dialog boxes describing how and what to do next, MINITAB persists as a popular choice for teaching students about statistics and data analysis. MINITAB primarily has a user base of educators using the program to show students research methods and analysis in college and graduate-level courses. MINITAB performs most general statistical analyses (regression, logistic regression, survival analysis, analysis of variance, factor analysis, but has its weaknesses in general linear model (GLM) and Multilevel regression).

e) STATA

It is a powerful statistical package with smart data-management facilities, a wide array of up-to-date statistical techniques, and an excellent system for producing publication-quality graphs. STATA latest version was produced June 24, 2013 which is a fast and easy to use data management package. STATA is available for Windows, Unix, and Mac computers. The standard version is called STATA/IC (or Intercooled STATA) and can handle up to 2,047 variables. There is a special edition called STATA/SE that can handle up to 32,766 variables (and also allows longer string variables and larger matrices), and a version for multi-core/multiprocessor computers called STATA/MP, which has the same limits but is substantially faster. STATA performs most general statistical analyses (regression, logistic regression, survival analysis, analysis of variance, factor analysis, multivariate analysis and time series analysis).

f) R and MATLAB

Stanford (2014) identified R and MATLAB as the richest statistical systems by far. They contain an impressive amount of libraries, which is growing each day. Even if a much desired specific model is not part of the standard functionality, you can implement it yourself, because R and MATLAB are really programming languages with relatively simple syntaxes. As "languages" they allow you to express any problem. In terms of modern applied statistics tools, R libraries are somewhat richer than those of MATLAB. R (http://cran.at.r-project.org) is a comprehensive statistical environment and programming language for professional data analysis and graphical display. It provides additional packages for statistical data analysis in different areas. The R software is free and runs on all common operating systems and it can be used to carry out some simple and complex analyses. On the flip side, MATLAB has much better graphics, which you will not be ashamed to put in a paper or a presentation. MATLAB and R perform most general statistical analyses (regression, logistic regression, survival analysis, analysis of variance, factor analysis, multivariate analysis). The greatest strengths of both are probably in its ANOVA, mixed model analysis and users creative freedom in analysis.

g) MS-EXCEL

Microsoft Excel is part of the Microsoft Office productivity suite has applications worldwide and one can use Excel to analyze data, for example, in accounts, budgets, billing and many other areas. Excel allows you to explore the menu bar and the different tasks that can be done with it. You can work on sample spreadsheets doing basic math, adding and deleting columns and rows, and preparing the worksheet for printing. You can run your data visually to show trends, patterns and comparisons between the data in a chart, table or other template, and Excel performs most general statistical analyses but weak in regression, logistic regression, survival analysis, analysis of variance, factor analysis, multivariate analysis.

h) ARTFISH

Approaches, Rules and Techniques for Fisheries statistical monitoring-ARTFISH has been developed as a standardized tool adaptable to most fisheries in the developing countries. Its design was driven by the need to provide users with robust, user-friendly and error-free approaches and computer software, and achieve the implementation of cost-effective fishery statistical systems with minimal external assistance.

i) BEAM 1 and 2

Bioeconomic Modeling of Artisanal and Industrial Sequential Shrimp Fisheries Software packages for bio economic modeling of artisanal and industrial sequential shrimp fisheries based on an age structured Thompson and Bell (in Ricker, 1975) yield per recruit biological model and a simple input output microeconomic model. BEAM1 gives its simulated results by age groups. BEAM2 gives them by standard commercial categories as used in the shrimp fishing industry.

j) BEAM 3

A Bio-economic Simulation Model of Tropical Shrimp Fisheries Using Fixed or Random Recruitment - BEAM3 is a stochastic model that aims at determining the optimum size, in the long term, of the level of fishing capacity corresponding to the management goals.

k) BEAM 4

Analytical Bio-economic Simulation of Space structured Multispecies and Multi fleets Fisheries – BEAM 4 is a versatile tool for the rational management of exploited living aquatic resources. It can deal with a fishery system of several stocks, several fleets, several areas (fishing grounds) and several processing plants and can account for migration of the animals as well as seasonality of recruitment.

l) CLIMPROD

Experimental Interactive software for choosing and fitting surplus production models including environmental variable - Various equations allow for the introduction of an environmental variable into surplus production models. CLIMPROD helps the user select the most appropriate model for a particular case, according to objective criteria. It resembles a simple expert system and uses artificial intelligence language (PROLOG) to converse with the user.

m) CLIMPROD-PLUS

Experimental Interactive software for choosing and fitting surplus production models including environmental variable - Various equations allowing for the introduction of an environmental variable into surplus production models are available. CLIMPROD helps the user to select the model corresponding to a particular case according to objective criteria. It looks like a simple expert-system and uses artificial intelligence to converse with the user.

n) FAST

Fishing Activity Simulation Tool-The Fishing Activity Simulation Tool, called FAST has been developed as an extension for the ESRI ArcView GIS Software (v.3.1) in combination with the Spatial Analyst v. 1.0 extension. It is based on two models focusing on the analysis of the spatial components of the fishing effort distribution.

o) FISAT II

Stock Assessment Tools - The FAO-ICLARM Stock Assessment Tools (FiSAT) is a program package consisting of robust methodologies for use with microcomputers, enabling users to formulate management options for fisheries, especially in data-sparse, tropical contexts.

p) FishStatJ

Software for fishery statistical time series-The system provides users with access to Fishery Statistics of various sorts. Any data having time series structure can potentially be stored and processed by FishStatJ. The system consists of a main module and the datasets. Currently one set of data is included within the set up module, while other sets are provided separately.

q) FishStat Plus

Universal software for fishery statistical time series-The system provides users with access to Fishery Statistics of various sorts. Any data having time series structure can potentially be stored and processed by FishStat Plus. The system consists of the main module and the datasets. Each dataset can be installed and uninstalled separately. This application, developed in 1998 (FishStat+), is NOT compatible with 64 bit operating system and Windows 7 and has now been replaced with FishStatJ.

r) MTBASE 1.1

The Model of Trophodynamics of Black - Azov Ecosystem - An ecology software package for the study of the impact of exotic species introduction on fish stocks: Mnemiopsis-Anchovy relationships case (MTBASE 1.1).

s) NANSIS

Software for Fishery Survey Data Logging and Analysis - NANSIS is a Survey Information System for logging, editing and analysis of scientific trawl survey data (trawl/catch data and length/frequency data).

t) SPATIAL

Space time Dynamics in Marine Fisheries-SPATIAL is a simulation package developed to model the space-time distribution of fishing intensity using alternative approaches.

u) Thompson and Bell Yield Analysis

This application refers to the use of a group of models loosely based on the work of Thompson and Bell (1934). The models are essentially mathematical depictions of fisheries, or parts of fisheries.

9.2 Features of Statistical Software

Statistical software has some common characteristics that make it reliable and suitable for data analysis: Data editor is in rows and columns which make it very easy to enter numeric data, There is availability of menu bar comprises drop-down menu, quick analysis as well as brief user manual, statistical level of measurement is put into consideration in data entry, They follow the initial steps in research project:

(a) Getting your data ready to enter into the software.
(b) Defining and labeling variable
(c) Entering data appropriately with each row containing each case and each column as variable.
(d) Data checking and cleaning is possible.

All data should be numeric, although it may not be all variables it is not desirable to use letter or word (String variable) as data. This can be achieved by recoding the letter or word (string data) into desirable numeric and labeled appropriately. Data exploration can be done to check for errors and other accuracy. The statistical level of significance for rejecting null hypothesis (Ho) is when your p-value significance is less than 0.05.

9.3 Common Statistical Tools and their Analysis by SPSS

Statistical analysis generally begins with the calculation of a number of summary statistics such as the mean, median, standard deviation, *etc.*, and by creating informative graphical displays of the data such as histograms, box plots, and stem-and-leaf plots. The aim at this stage is to describe the general distributional properties of the data, to identify any unusual observations (*outliers*) or any unusual patterns of observations that may cause problems for later analyses to be carried out on the data. Following the initial exploration of the data, statistical tests may be applied to answer specific questions or to test particular hypotheses about the data. According to Shamoo and Resnik (2003) various analytic procedures ¯provide a way of drawing inductive inferences from data and distinguishing the signal (the phenomenon of interest) from the noise (statistical fluctuations) present in the data.

The Statistical analysis depends on the objective of the study. The objective of a survey is to obtain information about the situation of the population study. The first Statistical task is therefore is to do a descriptive analysis of variables. In this analysis it is necessary to present results obtained for each type of variable. For qualitative and dichotomous variables, results must be presented as frequencies and percentages. For quantitative variables, the presentation is as means and deviations. After this analysis, you can access the association between variables and predictive analysis based on multiple regression models. An essential component of ensuring data integrity is the accurate and appropriate analysis of research findings. Improper statistical analyses distort scientific findings, mislead casual readers (Shepard, 2002), and may negatively influence the public perception of research. Integrity issues are just as relevant to analysis of non-statistical data as well. In deciding which test is appropriate to use, it is important to consider the type of variables that you have (*i.e.*, whether your variables are categorical, ordinal or interval and whether they are normally distributed. Some common statistical tools/techniques are furnished below:

Mean

The arithmetic mean, more commonly known as - the average is the sum of a list of numbers divided by the number of items on the list. The mean is useful in determining the overall trend of a data set or providing a rapid snapshot of your data. Another advantage of the mean is that it's very easy and quick to calculate.

Standard Deviation

The standard deviation, often represented with the Greek letter sigma, is the measure of a spread of data around the mean. A high standard deviation signifies that data is spread more widely from the mean, where a low standard deviation signals that more data align with the mean. In a portfolio of data analysis methods, the standard deviation is useful for quickly determining dispersion of data points.

Steps in SPSS

Enter data in SPSS Data Editor. There are two views in SPSS Data Editor. In variable view, one can define the name of variables and variable types such as string or numeric and data view gives the spreadsheet in which data pertaining to variables may be entered in respective columns. In the present case, we enter data in numeric format. Choose Analyze from the Menu Bar. Now select **Analyze → Descriptive Statistics → Frequencies....** Click on the **Statistics...** tab to get the Frequencies: Statistics dialog box. Select the desired options (Mean, Std. Deviation, Minimum, Maximum, Skewness and Kurtosis) required for analysis. Click on the **Continue** tab to return to the Frequencies dialog box. Click on **OK** to get the output.

Correlation and Regression

Correlation and Regression models the relationships between dependent and explanatory variables, which are usually charted on a scatter plot. The regression line also designates whether those relationships are strong or weak. Regression is commonly taught in high school or college statistics courses with applications for science or business in determining trends over time.

Steps in SPSS

Obtain correlation coefficient between each pair of the variables using the following steps. Once the data entry is complete, Choose Analyze from the Menu Bar. Now select **Analyze → Correlate → Bivariate.** For partial correlation use the following steps. Choose Analyze from the Menu Bar. Now select **Analyze → Correlate → Partial.**

For fitting a multiple linear regression equation by taking yield as dependent variable and biometrical characters (for example) as explanatory variables. Use the following steps: **Analyze → Regression → Linear.** In the Linear Regression **dialog** box select the Dependent Variable and the independent(s) variable box. For the Durbin-Watson statistics click **Statistics** on the Linear Regression dialog box. In the Linear Regression: Statistics dialog box check the Durbin-Watson check box under Residuals and other options as required. Click **Continue** to return to the Linear Regression dialog box. Click **Save** in the Linear Regression dialog box. In the Linear Regression: Save dialog box check the desired options under Predicted Values, Residuals, Distances and Influence Statistics. Click

Continue to return to the Linear Regression dialog box. In the Linear Regression dialog box Click **Options.** Click **Continue** to return to the Linear Regression dialog box. Click **OK.**

Hypothesis Testing

Also commonly called *t* testing, hypothesis testing assesses if a certain premise is actually true for your data set or population. In data analysis and statistics, you consider the result of a hypothesis test *statistically significant* if the results couldn't have happened by random chance. Hypothesis tests are used in everything from science and research to business and economic.

One Sample t-test

A one sample t-test allows us to test whether a sample mean (of a normally distributed interval variable) significantly differs from a hypothesized value *i.e.* the mean of the variable for this particular sample of students which is statistically (significantly) different from the test value. We would conclude that this group of students has a significantly higher mean on the writing test than the given.

Steps in SPSS: In the SPSS menu, select Analyze>Compare Means>One Sample t-test Select the variable(s) from the list you want to look at and click the button to move it into the "Test Variable(s)" area. Then enter the test value. In this example, we're testing the hypothesis that the median house value is 200,000. Select the Options button and check that the confidence interval is where you want it (the default is 95 per cent, which is what we normally use. Select "Continue" and then OK on the main window. You should get the following output.

Independent Samples T-test

This test is similar to the one-sample test, except rather than testing a hypothesized mean, we're testing to see if there is a difference between two groups. For the grouping variable, you can choose a demographic trait (such as gender, age, ethnicity, *etc.*) or any other variable that classifies your groups. (In an experimental design, it is a good way to test the differences between the control group and the manipulation group.) In this example, we'll use gender.

Steps in SPSS

In the SPSS menu, select Analyze>Compare Means>Independent Samples T-test Select your Test variable from the list. This is the variable for which you want to compare means. Now select the grouping variable, which is the trait you're using to divide the groups. For this example, we will select gender. Select it from the list and click the arrow next to "Grouping Variable." Then, click "Define Groups ..." and enter the values for the two groups. In this example, 1=Female and 2=Male. Also notice the "Cut Point" option. What if we wanted to divide the sample into two groups based on home value? The cut point would be the value where you split the sample. For example, if you entered 100,000, it would create two groups – one for home value less than 100,000 and another for more than 200,000. For this example, let's stick to gender, though. Select continue, and then click "OK."

Paired Samples t-test

With the paired samples t-test, we're not testing for differences between groups. Instead, we're testing for means of different variables within the sample. For example, we want to compare the mean for user-created videos and the mean for company generated videos. Go to Analyze>Compare Means>Paired Samples T-test Select the two variables you want to compare, and click the arrow to move them into the "Paired Variables" pane. Under options, make sure that you're using a 95 per cent confidence interval. Clicks continue and then OK and you'll get the following output.

One-Way ANOVA

What if we want to test for differences between more than two groups. ANOVA (which stands for "Analysis of Variance") is the way to go. Analysis>Compare Means>One-Way ANOVA Select the variable(s) you want to test and move into the "Dependent List" pane. Now, move the variable that you are using to separate them into groups into the "Factor" pane. For example, in our data set, there are three age groups ("1","2" and "3") for the 15-18, 19-24, and 25+ groups, respectively, in the Age Group variable. Click "OK" and you get the following output. To determine which one of the means is significantly different, we must use a post-hoc test. Go back to Analyze>Compare Means>One Way ANOVA but this time, click Post-Hoc, Then, select "Tukey" and "Duncan".

One Sample Median Test

A one sample median test allows us to test whether a sample median differs significantly from a hypothesized value. The one-sample median test is used to test a hypothesized median value against an observed median value in a representative sample.

For instance, let's say researchers believe that the college-age students will rate the look of a new social networking site at as a "4" on a 5-point Likert scale. Researchers collect a representative sample from a Psychology 101 class and find a median value of "1." This difference of ratings between what researchers expected (Median = 4) and the observed value (Median = 1) may be statistically significant according to a one-sample median test. The data is entered in a between-subjects fashion and is specified as a **scale** variable in the **Variable View**. Click **Analyze.** Drag the cursor over the **Nonparametric Tests** drop-down menu. Click **One Sample**. Click on the **Settings** tab. Under the **Customize tests marker**, click on the **Compare median to hypothesized (Wilcoxon Signed-**

rank test) box to select it. In the **Hypothesized median:** box, type in the *a priori* **hypothesized median value**. Click on the **Run** button.

Binomial Test

A one sample binomial test allows us to test whether the proportion of successes on a two-level categorical dependent variable significantly differs from a hypothesized value. Go to Analyze > Nonparametric Tests > Legacy Dialogue > Binomial Test

Chi-square Goodness of Fit

A chi-square goodness of fit test allows us to test whether the observed proportions for a categorical variable differ from hypothesized proportions.

First Step

Select Data → Weight Cases. A box pops up with a list of the input variables. Click the "Weight cases by" button, and then move the count variable into the box under "Frequency Variable:" Finally, click "OK."

Second Step

Select Analyze → Nonparametric Tests → Chi-Square. Move the variable indicating categories into the "Test Variable List:" box. If you want to test the hypothesis that all categories are equally likely, click "OK." Otherwise, categories should be labeled in ascending numerical order, say 1, 2, 3,., and you must specify the expected count for each category by first clicking the "Values" button under "Expected Values." Then, in the box to the right of "Values," enter the expected count for category 1 and click the "Add" button. Now enter the expected count for category 2 and click "Add." Continue in this way until all expected counts have been entered. Then click "OK."

Notes

(1) By default SPSS assumes that the chi-square degrees of freedom are k-1, where k is the number of categories. If your null hypothesis does not completely specify all probabilities, then you will have to determine the correct P-value yourself. (2) SPSS will not perform a chi-square test if one or more observed counts are 0. In this case, use other software or combine categories.

Statistical software packages have been recognized to have contributed immensely to research analysis by helping to minimizing human and experimental errors in data analysis. Conclusively, it is the opinion of the researcher, after analyzing the field data to a very large extent, that statistical software has contributed immensely to social research especially in the area of demographic and data analysis. This was achieved by using a scientific approach to solve fundamental problems in research which is data analysis.

1. It has been discovered that some analysis such as post Hoc, complex analysis in time series, regression and variance analysis cannot be calculated manually effectively without statistical software packages.
2. Statistical software packages have been discovered to help academic staffs in higher institution to improve their research expertise by attending training on usage of statistical packages.
3. Statistical packages make research work robust and faster.
4. It was discovered that maximum efficiency of researcher in statistical software is determine by the years of experience in usage and the area of specialization.
5. Most reason for using statistical software was it easy usage, suitability for many statistical analysis and the likes.
6. While reason for non usage range from lack of attention to learn, difficult usage, cost of licensing, *etc.* It was discover that statistical software are not expensive neither are they too difficult to use but people do not give attention to its learning.

REFERENCES

Adetola O.G (2013) Learning Statistical Package Workbook. Nigeria (Unpublised).

Akindutire, A.F. (2013) Usage of Mobile Phone: Deception, Deviance and Fraud (A case study of Ambrose-Alli) Chris McGann, Role of Statistical Software, eHow Contributor http://www.ehow.com/facts_6923793_role-statistical-software.html#ixzz2xoZFcTZX visited on 3rd April, 2014.

Daniel, S. (2009). *The Most Popular Statistical Software.* accessed on 3rd April, 2014: eHow Contribution Eview Users Manual (2009) History and usage of Eview visited on 28th April, 2014

Gottschalk, L. A. (1995). Content analysis of verbal behavior: New findings and clinical applications. Hillside, NJ: Lawrence Erlbaum Associates, Inc International Organization of Scientific Research, IOSR.

Karen (2013) Choosing a Statistical Software Package or Two Source:http://www.theanalysisfactor.com/choosing-statistical-software/VISITED ON 7TH April, 2014

McCullough, B.D. (1999). "Econometric software reliability: EViews, LIMDEP, SHAZAM and TSP". Journal of Applied Econometrics 14 (2):191–202. HYPERLINK

McGann, C. (2009). Role of Statistical Software. eHow Contribution accessed by 3rd April, 2014

Resnik, D. (2000). Statistics, ethics, and research: an agenda for educations and reform. Accountability in Research. 8: 163-88

Savenye, Robinson, 2004 Clinical significance of research: A growing concern. Canadian Journal of Nursing Research, 24, 1-4.

Schroder, K.E., Carey, M.P., Venable, P.A. (2003). Methodological challenges in research on sexual risk behavior: I. Item content, scaling, and data analytic options. Ann Behav Med, 26(2): 76-103.

Shamoo, A.E., Resnik, B.R. (2003). Responsible Conduct of Research. Oxford University Press.

Silverman, S., Manson, M. (2003). Research on teaching in physical education doctoral dissertations: a detailed investigation of focus, method, and analysis. Journal of Teaching in Physical Education, 22(3): 280-297.

Smeeton, N., Goda, D. (2003). Conducting and presenting social work research: some basic statistical considerations. Br J Soc Work, 33: 567-573.

STATA PRESS site (2014) Stata history and usage.visited 28th April, 2014.

W.N. Venables and B.D. Ripley. Modern Applied Statistics with S. Springer, first edition, 1994.

W.N. Venables and B.D. Ripley. S Programming. Statistics and Computing. Springer Verlag, New York, N.Y., 2000.

Zaslavsky Alan (2014) HYPERLINK "http://www.hcp.med.harvard.edu/statistics/survey-soft/Summary of Survey Analysis Software wikipedia library

— Chapter 10 —

Grounded Theory

A study using grounded theory is likely to begin with a question, or even just with the collection of qualitative data-As researchers review the data collected, repeated ideas, concepts or elements become apparent, and are tagged with codes, which have been extracted from the data. As more data is collected, and re-reviewed, codes can be grouped into concepts, and then into categories. These categories may become the basis for new theory. Thus, grounded theory is quite different from the traditional model of research, where the researcher chooses an existing theoretical framework, and only then collects data to show how the theory does or does not apply to the phenomenon under study.

Grounded theory is a general methodology, a way of thinking about and conceptualizing data. It focuses on the studies of diverse populations from areas like remarriage after divorce (Cauhape, 1983) and Professional Socialization (Broadhed, 1983). The grounded theory method was developed by two sociologists'. Barney Glaser and Anselm Strauss.'

1. Rationale of the theory to be grounded is that this theory helps close the gap between theory and empirical research.
2. Helped in suggesting the logic of grounded theories.
3. This book helped to legitimize careful qualitative research. This was seen to be the most important goal because, by the 1960s, quantitative research methods had taken an upper hand in the fields of research and qualitative methods were not seen as adequate methods of verification.

Introduction

The methodology of grounded theory was developed by American sociologists Glaser and Strauss in 1967 to describe a new qualitative research method they used in their research Awareness of Dying in 1965. In this study, they adopted an investigative research method with no preconceived hypothesis and used continually comparative analysis of data. They believe that the theory obtained by this method is truly grounded in the data. For this reason they named the methodology "grounded theory" (Glaser and Strauss, 1967).

The goal of the grounded theory approach is to generate a theory that explains how an aspect of the social world "works". The goal is to develop a theory that emerges from and is therefore connected to the very reality that the theory is developed to explain.

Key Definitions

According to Creswell (2009), grounded theory is "a qualitative strategy of inquiry in which the researcher derives a general, abstract theory of process, action, or interaction grounded in the views of participants in a study." (p. 13 and 229) this process involves using multiple stages of data collection and the refinement and interrelationships of categories of information (Charmaz, 2006, Strauss and Corbin, 1990, 1998).

Other Definitions of Grounded Theory

Grounded theory is "a systematic qualitative research methodology in the social sciences emphasizing generation of theory from data in the process of conducting research." (Martin, *et al.*, 1986) "The grounded theory approach is a qualitative research method that uses a systematic set of procedures to develop an inductively derived grounded theory about a phenomenon." (Strauss and Corbin, 1990)

A complete grounded theory research design often contains the elements listed in Table 1. These steps may not be undertaken sequentially in the research; the researchers sometimes need to go back and forth amongst several steps.

Table 1: General Elements in a Grounded Theory Research Design

1.	Question formulating
2.	Theoretical sampling
3.	Interview transcribing and Contact summary
4.	Data chunking and Data naming - Coding
5.	Developing conceptual categories
6.	Constant comparison
7.	Analytic memoing
8.	Growing theories

Defining Features

Two primary characteristics of grounded theory research design:

1. The constant comparison of data with emerging categories and,
2. Theoretical sampling of different groups to maximize the similarities and differences of information (CreswelL 2009, p.13).

Current Uses of Grounded Theory

Grounded theory is a powerful research method for collecting and analyzing data. Traditional research designs which usually rely on a literature review leading to the formation of a hypothesis. Then one tests the hypothesis through experimentation in the real world.

Grounded theory investigates the actualities in the real world and analyses the data with no preconceived ideas or hypothesis (Glaser and Strauss, 1967). In other words, grounded theory suggests that theory emerges inductively from the data (Chesebro and Borisoff, 2007). Though it can be used in different types of research, grounded theory is often adopted to formulate hypotheses or theories based on existing phenomena, or to discover the participants' main concern and how they continually try to resolve it (Glaser, 1992).

Strengths and Weaknesses

Due to the difficulties and weaknesses encountered when applying grounded theory, this methodology is still not widely used or understood by researchers in many disciplines (Allan, 2003).

Strengths

- ☆ An effective approach to build new theories and understand new phenomena.
- ☆ High quality of the emergent theory.
- ☆ Emergent research design reflects the idiosyncratic nature of the study.
- ☆ Findings and methods are always refined and negotiated.
- ☆ Requires detailed and systematic procedures for data collection, analysis and theorizing.
- ☆ The resulting theory and hypotheses help generate future investigation into the phenomenon.
- ☆ Requires the researcher to be open minded, and able to look at the data through many lenses.
- ☆ Data collection occurs over time, and at many levels, helping to ensure meaningful results.

Weaknesses

- ✰ Huge volumes of data.
- ✰ Time consuming and painstakingly precise process of data collection/ analysis.
- ✰ Lots of noise and chaos in the data.
- ✰ Prescribed application required for the data-gathering process.
- ✰ There are tensions between the evolving and inductive style of a flexible study and the systematic approach of grounded theory.
- ✰ It may be difficult in practice to decide when the categories are "saturated" or when the theory is sufficiently developed.
- ✰ It is not possible to start a research study without some pre- existing theoretical ideas and assumptions.
- ✰ Requires high levels of experience, patience and acumen on the part of the researcher.

Data Collection

This is not to suggest that classic grounded theory is free of any theoretical lens but rather that it should not be confined to any one lens; that as a general methodology, classic grounded theory can adopt any epistemological perspective appropriate to the data and the ontological stance of the researcher (Holton, 2009).

Data collection of grounded theory is directed by theoretical sampling, which means that the sampling is based on theoretically relevant constructs, it enables the researcher to select subjects that maximize the potential to discover as many dimensions and conditions related to the phenomenon as possible (Strauss and Corbin, 1998).

Many experiments, in their early stages, use the open sampling methods of identifying individuals, objects or documents. This is so that the data's relevance to the research question can be assessed early on, before too much time and money has been invested (Davidson, 2002).

Grounded theory data collection is usually but not exclusively by interviews. Actually, any data collection method can be used, like focus groups, observations, informal conversation, group feedback analysis, or any other individual or group activity which yields data (Dick, 2005).

Interview transcribing is probably one of the most time-consuming parts of the research. The researchers are suggested to transform the tape recordings of interviews and other notes into word-by-word transcripts for further analysis. However, some researchers (Glaser, 1992, Dick, 2005) argue that taking key-word notes during the interviews, tape-recording the interviews and checking the notes against the tape recording and converting them to themes afterwards can also do the job well, and is less time-consuming.

Data Analysis and Interpretation

I believe grounded theory draws from literary analysis, and one can see it here. The advice for building theory parallels advice for writing a story. Selective coding is about finding the driver that impels the story forward. (Borgatti).

Grounded theory data analysis involves searching out the concepts behind the actualities by looking for codes, then concepts and finally categories.

1. Codes

Coding is a form of content analysis to find and conceptualize the underlying issues amongst the 'noise" in the data. During the analysis of an interview, the researcher will become aware that the interviewee is using words and phrases that highlight an issue of importance or interest to the research; they are noted and described in a short phrase. The issue may be mentioned again in the same or similar words and is again noted. This process is called coding and the short descriptor phrase is a code (Allan, 2003).

Example

"Pain relief is a major problem when you have arthritis. Sometimes, the pain is worse than other times, but when it gets really bad, whew! It hurts so bad, you don't want to get out of bed. You don't feel like doing anything. Any relief you get from drugs that you take is only temporary or partial." (Interviewee).

One thing that is being discussed here is PAIN. Implied in the text is that the speaker views pain as having certain properties, one of which is INTENSITY: it varies from a little to a lot. (When is it a lot and when is it little?) When it hurts a lot, there are consequences: don't want to get out of bed, don't feel like doing things (what are other things you don't do when in pain?). In order to solve this problem, you need PAIN RELIEF. One AGENT OF PAIN RELIEF is drugs (what are other members of this category?) Pain relief has a certain DURATION (could be temporary), and EFFECTIVENESS (could be partial).

Coding Procedures in Grounded Theory Approaches

Strauss and Corbin (1990) describe some flexible guidelines for coding data when engaging in a Grounded Theory analysis:

Open Coding: form initial categories of information about the phenomenon being studied from the data gathered. This is "the process of breaking down, examining, comparing, conceptualizing, and categorizing data' 1 (p. 61).

Axial Coding: involves assembling the data in new ways after open coding. A coding paradigm (logic diagram) is then developed which:

- ☆ Identifies a central phenomenon.
- ☆ Explores causal conditions.
- ☆ Identifies the context and intervening conditions.

- ✰ Specifies strategies.
- ✰ Delineates the consequences.

Selective Coding: involves the integration of the categories in the axial coding model. In this phase, conditional propositions (or hypotheses) are typically presented. The result of this process of data collection and analysis is a substantive-level theory relevant to a specific problem, issue or group. It is "the process of selecting the core category, systematically relating it to other categories, validating those relationships, and filling in categories that need further refinement and development' 1 (p. 116).

Note 1: The three types of coding are not necessarily sequential; they are likely to overlap. After collecting additional data, the researchers return to analyzing and coding data, and use the insights from that analysis process to inform the next iteration of data collection. This process continues until a strong theoretical understanding of an event, object, setting or phenomenon has emerged. (Constant Comparative Method).

Note 2: As mentioned, the process of naming or labeling objects, categories, and properties is known as coding. Coding can be done very formally and systematically or informally. In grounded theory, it is normally done quite informally- For example, if after coding much text, some new categories are invented; grounded theorists do not normally go back to the earlier text to code for that category. However, maintaining an inventory of codes with their descriptions (*i.e.*, creating a codebook) is useful, along with pointers to text that contain them. In addition, as codes are developed, it is useful to write memos known as code notes that discuss the codes. These memos become fodder for later development into reports.

2. Concepts

Codes are then analyzed and those that relate to a common theme are grouped together. This higher order commonality is called a concept (Allan, 2003).

Note: based on our understanding, the process of inducting concepts is overlapping with the three types of coding process (mentioned above). They are basically the same but different researchers give them different descriptions according to their specific research experience.

3. Categories

Concepts are then grouped and regrouped to find yet higher order commonalities called categories. It is these concepts and categories that lead to the emergence of a theory (Allan, 2003).

"An effective strategy is. at first, literally to ignore the literature of theory and fact on the area under study, in order to assure that the emergence of categories will not be contaminated by concepts more suited to different areas." (Glaser and Strauss, 1967).

Note: according to Strauss and Corbin (1998), grounded theory has particular types of prescribed categories as components of the theory. But this may not appear appropriate for a particular study.

To Recap: developing a grounded theory model involves systematically analyzing a phenomenon in order to explain how the process occurs inductively (Strauss and Corbin, 1998).

Standards of Validation

Strauss and Corbin (1990) state that there are four primary requirements for judging grounded theory:

1. It should fit the phenomenon, provided it has been carefully derived from diverse data and is adherent to the common reality of the area;
2. It should provide understanding, and be understandable;
3. Because the data is comprehensive, it should provide generality, in that the theory includes extensive variation and is abstract enough to be applicable to a wide variety of contexts; and
4. It should provide control, in the sense of stating the conditions under which the theory applies and describing a reasonable basis for action.

Grounded theory is about adopting a constant comparative method, therefore the conformity and coherence of codes, concepts and categories is also an important indicator for a valid grounded theory. This means that a grounded theory is reliable when there comes no new categories in the data collected. This means one can say the theory is sufficiently developed.

The process under which the theory has been developed can evaluate the quality of a theory. This contrasts with the scientific perspective that how you generate a theory is not as important as its ability to explain new data.

The researcher should not switch their focus from abstraction to description as concepts emerge. Detailed description offers data for conceptual abstraction and the possible emergence of a grounded theory in the future, but cannot be considered grounded theory.

Deciding to use grounded theory means embracing it fully (not pieces of it) It requires the adoption of a systematic set of precise procedures for collection, analysis and articulation of conceptually abstract theory.

Report Writing and Rhetorical Structure

Glaser and Strauss (1967) describe 4 main stages in building grounded theory:

1. Comparing Incidents Applicable to each Category

Begin by coding the data into as many categories as possible. Some categories will be generated by the researcher and some from the language and data of the research situation. As more instances of the same category code are found ideas about that category can be refined. At this point it's best to stop coding and make a memo of these ideas.

2. Integrating Categories and their Properties

The constant comparative method will begin to evolve from comparing incidents to focusing on emergent properties of the category. Diverse properties will start to become integrated. The resulting theory will begin to emerge by itself.

3. Delimiting the Theory

Eventually the theory comes together, and there are fewer changes to the theory as the researcher compares more incidents. Later modifications include taking out irrelevant properties of categories, and adding details of properties into an outline of interrelated categories. More importantly, the researcher begins to find ways to delimit the theory with a set of higher level concepts. The researcher needs to generalize the theory more as they continue to make constant comparisons against it. The number of categories will be reduced.

New categories are often created halfway through coding and it usually isn't necessary to go back and code for them. The researcher only needs to code enough to saturate the properties of the category. Later the researcher can evaluate the categories and emergent theory by moving on to new comparison groups.

4. Writing Theory

"When the researcher is convinced that his analytic framework form a systematic substantive theory, that it is a reasonably accurate statement of the matters studied, and that it is couched in a form that others going into the same field could use - then he can publish his results with confidence".

A Review of Study: Qualitative Tussles in Undertaking a Grounded Theory Study Purpose

This paper, by Judith A. Holton, is a methodological critique of Classic Grounded Theory (as developed by Glaser). Holton attempts to identify and clarify some of the key misconceptions in the use and understanding of Grounded Theory. She uses examples of research studies that have been performed under the guise of grounded theory, but are only using fragments of the grounded theory methodology. Holton explains how this does not constitute true grounded theory research.

Key Points

Personal bias: grounded theory literature often states the need to have no preconceived notions or frameworks in mind when conducting the research. It seems impossible to ignore ones worldview (and it is). The point is to be able to look at the phenomenon and emerging data from many lenses.

The data fit: one of the biggest problems (as seen by classic grounded theorists) is when researchers dismiss data altogether because it does not "fit". In grounded theory the data that does not fit established theories and frameworks is the important data! This is what will lead to a totally new view/interpretation of the phenomenon under study.

- ☆ Giving in: There is a tendency for researchers who undertake grounded theory to fold, or become lenient in their application of the rigid and time consuming process of data analysis. Grounded theory is time consuming and often frustrating. This must be understood and embraced if the process is to be successful.
- ☆ Description vs. explanation: Explanation of patterns of behaviour is the ultimate goal of grounded theory research. Description of what is happening is often seen as a substitution. These two outcomes are not interchangeable. It is not about accuracy of description, it is about conceptual abstraction, resulting in conceptual hypotheses.
- ☆ Role of context: The context of the study should not influence data analysis from the outset. The context should be seen as another piece of the puzzle that may or may not be of importance. If it is of importance this will emerge naturally from the participants.

Checklist

- ☆ What is the phenomenon of interest?
- ☆ Does grounded theory best suit the study of the phenomenon?
- ☆ Is there existing literature on the specific area of interest?
- ☆ Are there theories that adequately explain the occurrences within the phenomenon?
- ☆ What is the role of the researcher in the study?
- ☆ Is the body of literature acting as additional data?
- ☆ Is it ensured the context does not influence data analysis?
- ☆ What is the researcher's relationship to the study?
- ☆ What precautions will be taken to ensure unbiased approach of the researcher?
- ☆ How will constant comparative analysis occur?
- ☆ Who are the subjects of interest?

- What is the data collection method?
- What are the coding procedures?
- How will relationships between concepts be identified and categorized?
- Are the results new explanations of relationships?
- Is the process constantly reflexive?

Conclusions and Recommendations

The value of grounded theory is in its ability to examine relationships and behaviour within a phenomenon from an unbiased in-depth perspective. That is to say, when a researcher enters a study with no framework or theory they are wish to fit the data into the doors are open to discovering explanations that have yet to be articulated. More importantly, the explanations ultimately come from the participants being studied. When a grounded theory study is executed correctly and rigorously, there is little chance that the resulting explanations have distorted by the researchers personal worldview.

The time and detailed analysis required to properly execute grounded theory methodology makes its use daunting and limited There are many variables that must be in place (*i.e.* resources, experience of researcher, acceptance of methodological processes *etc.*) in order for grounded theory to be successfully carried out. When this occurs the results can be invaluable to the understanding of social phenomena.

REFERENCES

Allan, G. (2003). A critique of using grounded theory as a research method.

Electronic Journal of Business Research Methods, 2(1).

Charmaz, K. (2006). Constructing grounded theory. Thousand Oaks, CA: Sage.

Chesebro, J.W., and Borisoff, D.J. (2007). What makes qualitative research qualitative? Qualitative Research Reports in Communication- 8(1): 3-14

Creswell, J.W. (2009). Research Design: Qualitative, Quantitative, and Mixed Approaches. Thousand Oaks, CA: Sage.

Dick, B. (2005). Grounded theory: a thumbnail sketch. [On line] Available at http://www.scu.edu.au/schools/gcm/ar/arp/grounded.html

Glaser, B. (1992). Basics of grounded theory analysis. Mill Valley, CA' Sociology Press.

Glaser. B.G. and Strauss, A.L. (1967). The Discovery of Grounded Theory.

Strategies for Qualitative Research. Aldine Transaction, Inc.

Holton, J.A. (2009). Qualitative Tussles in Undertaking a Grounded Theory Study The Grounded Theory Review, 8(3): 37-49.

Martin, *et al.* (1986). Grounded Theory and Organizational Research. The Journal of Applied Behavioral Science, 22(2): 141.

Strauss, A. and Corbin, J. (1990). Basics of qualitative research: Grounded theory procedures and techniques (1st ed.), Newbury Park, CA: Sage.

Strauss, A., and Corbin; J. (1998). Basics of qualitative research: Grounded theory procedures and techniques (2nd ed.). Newbury Park, CA: Sage.

Index

T

V

W

9 789359 191744

Printed at Repro India Ltd.